EAI/Springer Innovations in Communication and Computing

Series Editor

Imrich Chlamtac, European Alliance for Innovation, Ghent, Belgium

Editor's Note

The impact of information technologies is creating a new world yet not fully understood. The extent and speed of economic, life style and social changes already perceived in everyday life is hard to estimate without understanding the technological driving forces behind it. This series presents contributed volumes featuring the latest research and development in the various information engineering technologies that play a key role in this process. The range of topics, focusing primarily on communications and computing engineering include, but are not limited to, wireless networks; mobile communication; design and learning; gaming; interaction; e-health and pervasive healthcare; energy management; smart grids; internet of things; cognitive radio networks; computation; cloud computing; ubiquitous connectivity, and in mode general smart living, smart cities, Internet of Things and more. The series publishes a combination of expanded papers selected from hosted and sponsored European Alliance for Innovation (EAI) conferences that present cutting edge, global research as well as provide new perspectives on traditional related engineering fields. This content, complemented with open calls for contribution of book titles and individual chapters, together maintain Springer's and EAI's high standards of academic excellence. The audience for the books consists of researchers, industry professionals, advanced level students as well as practitioners in related fields of activity include information and communication specialists, security experts, economists, urban planners, doctors, and in general representatives in all those walks of life affected ad contributing to the information revolution.

Indexing: This series is indexed in Scopus, Ei Compendex, and zbMATH.

About EAI

EAI is a grassroots member organization initiated through cooperation between businesses, public, private and government organizations to address the global challenges of Europe's future competitiveness and link the European Research community with its counterparts around the globe. EAI reaches out to hundreds of thousands of individual subscribers on all continents and collaborates with an institutional member base including Fortune 500 companies, government organizations, and educational institutions, provide a free research and innovation platform. Through its open free membership model EAI promotes a new research and innovation culture based on collaboration, connectivity and recognition of excellence by community.

Anandakumar Haldorai • Arulmurugan Ramu •
Sudha Mohanram • Rita Zgheib
Editors

4th EAI International Conference on Big Data Innovation for Sustainable Cognitive Computing

BDCC 2021

Springer

EAI
RESEARCH MEETS INNOVATION

Editors
Anandakumar Haldorai
Department of Computer Science and Engineering
Sri Eshwar College of Engineering
Coimbatore, Tamil Nadu, India

Arulmurugan Ramu
Department of Computer Science and Engineering
CMR University
Bengaluru, KA, India

Sudha Mohanram
Sri Eshwar College of Engineering
Coimbatore, Tamil Nadu, India

Rita Zgheib
Canadian University Dubai
Dubai, United Arab Emirates

ISSN 2522-8595 ISSN 2522-8609 (electronic)
EAI/Springer Innovations in Communication and Computing
ISBN 978-3-031-07653-4 ISBN 978-3-031-07654-1 (eBook)
https://doi.org/10.1007/978-3-031-07654-1

This Springer imprint is published by the registered company Springer Nature Switzerland AG
The registered company address is: Gewerbestrasse 11, 6330 Cham, Switzerland

Preface

We are delighted to introduce the proceedings of the fourth edition of 2021 European Alliance for Innovation (EAI) International Conference on Big Data Innovation for Sustainable Cognitive Computing (BDCC 2021). This conference has brought together researchers, developers, and practitioners around the world who are leveraging and developing big data technology for a smarter and more resilient data computing.

The technical program of BDCC 2021 consisted of 18 full papers under oral presentation session at the main conference tracks. Aside from the high-quality technical paper presentations, the technical program also featured keynote designed for professionals working in the early stages of building an advancement program, as well as those with more mature operations. The keynote was Dr. Rita Zgheib, Department of Computer Engineering and Computational Science School of Engineering, Applied Science and Technology, Canadian University Dubai, United Arab Emirates.

Coordination with the steering chair Dr. Imrich Chlamtac and Dr. Anandakumar Haldorai was essential for the success of the conference. We sincerely appreciate their constant support and guidance. It was also a great pleasure to work with such an excellent organizing committee for their hard work in organizing and supporting the conference. In particular, the Technical Program Committee, led by our TPC Chair Dr. Arulmurugan Ramu, Publication Chair Prof. M. Suriya, and Local Committee Chairs Prof. K. Karthikeyan and Dr. K. Aravindhan, completed the peer-review process of technical papers and made a high-quality technical program.

We are also grateful to our Conference Manager Ms. Karolina Marcinova and Publication and Managing Editor Ms. Eliska Vlckova for their continuous support and guidance. We thank all the authors who submitted and presented their papers to the BDCC 2021 conference and workshops.

We strongly believe that BDCC 2021 conference has provided a good forum for all researcher, developers, and practitioners to discuss all science and technology aspects that are relevant to big data technology. We also expect that the future BDCC 2022 conference will be as successful and stimulating, as indicated by the contributions presented in this series.

Coimbatore, Tamil Nadu, India Anandakumar Haldorai

Bengaluru, KA, India Arulmurugan Ramu

Coimbatore, Tamil Nadu, India Sudha Mohanram

Dubai, United Arab Emirates Rita Zgheib

Contents

Part III Bigdata Emerging Applications

Part IV Bigdata and Technology

Part V Bigdata in Medical Applications

Conference Organization

Steering Committee

Imrich Chlamtac Bruno Kessler Professor, University of Trento, Trento, Italy
Sudha Mohanram Sri Eshwar College of Engineering, Coimbatore, India
Anandakumar Haldorai Sri Eshwar College of Engineering, Coimbatore, Tamil Nadu, India
Arulmurugan Ramu CMR University, Bengaluru, India

Organizing Committee

General Chair

Anandakumar Haldorai Sri Eshwar College of Engineering, Coimbatore, Tamil Nadu, India

TPC Chair

Arulmurugan Ramu CMR University, Bengaluru, India

Sponsorship and Exhibit Chair

Michaelraj Kingston Roberts Sri Eshwar College of Engineering, Coimbatore, India

Local Chair

K. Sivakumar National Institute of Technology, Karnataka, India

Workshops Chair

K. Aravindhan SNS College of Engineering, Coimbatore, India

Publicity & Social Media Chair

V. S. Akshaya Sri Eshwar College of Engineering, Coimbatore, India

Publications Chair

Suriya Murugan Sri Eshwar College of Engineering, Coimbatore, India

Web Chair

K. Karthikeyan SNS College of Engineering, Coimbatore, India

Conference Manager

Karolina Marcinova EAI, Gent, Belgium

Technical Program Committee

K. Aravindhan SNS College of Engineering, Coimbatore, India
M. Suriya Vel Tech Rangarajan Dr. Sangunthala R&D Institute of Sci and Tech, Chennai, India
A. Anandaraj Sri Eshwar College of Engineering, Coimbatore, India
Abildtrup Abdulhaq Roskilde University, Roskilde, Denmark
Matt Bowden Australian National University, Canberra, ACT, Australia
Debabrata Datta GLA University, Mathura, Uttar Pradesh, India
A. S. Deepalakshmi Maryland University, MD, USA
Sim Sze Yin Diana National Institute of Technology, Singapore, Singapore
Folsom Duvalf University of Oklahoma, Norman, OK, USA
Jain Emadi McMaster University, Hamilton, ON, USA
Karthikeyan Kandasamy Anna University, Chennai, India
S. Saranya Dr. NGP Institute of Technology, Coimbatore, India
S. Shanthi Sri Eshwar College of Engineering, Coimbatore, India
Lin Shih-Chau National Ocean University, Keelung, Taiwan
Lee Su K Young Yonsei University, Seoul, Korea

Part I
Bigdata Services

Performance Evaluation of Adaptive Neuro Fuzzy Inference System (ANFIS) for the Prediction of Cloud Service Provider

M. Sujatha and K. Geetha

Abstract In the current scenario, cloud adoption has become inevitable due to reduced infrastructure cost, increased scalability, and less maintenance. Many cloud service providers (CSP) are emerging to satisfy this rapid demand and have heterogeneous service plans with complex pricing policies. This makes the cloud service selection process a daunting task for the customer. User preference of QoS parameters is an important factor in deciding the optimal CSP, but uncertainty in the judgment of these parameters makes it a viable candidate for applying a fuzzy inference system. We have analyzed and implemented the adaptive neural fuzzy inference system (ANFIS) on QWS dataset for the prediction of CSP. We have applied subtractive clustering and compared with fuzzy c-means clustering to reduce the error rate in the prediction. We have done statistical analysis using different metrics such as RMSE, squared R, mean absolute error (MAE), and combined accuracy(CA) to measure the performance of the model and proved its efficiency.

Keywords ANFIS · QoS attribute · Fuzzy inference system · Cloud computing · CSP selection

1 Introduction

Adoption to cloud services is becoming more popular all over the world due to the pandemic situation. Many small enterprises and start-ups prefer cloud services to support their infrastructure demand that in turn reduces the initial investment. Also, the dynamic resource demand of each application is satisfied with the cloud because

M. Sujatha (✉) · K. Geetha
Sastra Deemed to be University, Thanjavur, Tamil Nadu, India
e-mail: sujatha@cse.sastra.edu; geetha@cse.sastra.edu

A. Haldorai et al. (eds.), *4th EAI International Conference on Big Data Innovation for Sustainable Cognitive Computing*, EAI/Springer Innovations in Communication and Computing, https://doi.org/10.1007/978-3-031-07654-1_1

of autoscaling and elasticity. According to the latest forecast from Gartner Inc.,[1] worldwide end-user spending on public cloud services is forecast to grow 23.1% in 2021 to total $332.3 billion, up from $270 billion in 2020. Due to this increase in demand for cloud services, many vendors have entered into this market with a wide variety of service offerings. Each application has specific requirements in terms of response time, availability, security, etc., and so before moving into the cloud, select the optimal CSP to satisfy these specific requirements. Many vendors are delivering services with similar quality of service and performance, so understanding these services and choosing the optimal service provider have become a tedious task for the end user. Many researches have been conducted based on QoS parameters to improve the accuracy of selection as well as distinguishing the performance of various service providers. Some of the QoS parameters are qualitative, and some are quantitative. Many of the providers are not disclosing these data in public, and so the decision of optimal service by the user becomes a difficult task. Consideration of these parameters is essential because the reliability of the providers is proved by mapping with committed and actual performance.

The selection process involves consideration of several criteria and multiple alternatives, and so many researches have considered it as an MCDM problem. Many of the criteria for the selection process is not transparently given by the provider, and so the user requirements will be of uncertain nature that can be represented as a fuzzy set.

Main contribution of our work is as follows:

1. The different clustering techniques are evaluated to construct the knowledge base.
2. The fuzzy model is automatically tuned at different epochs, variable step size, and optimization methods to reduce the error rate.
3. Performance of the model is evaluated by RMSE (root mean square error) and squared R.

The rest of this chapter is organized as follows: Sect. 2 describes the existing work related to cloud service selection problem and ANFIS model. Section 3 provides the background study and the proposed methodology for the experiment. Section 4 is about the implementation of this model with QWS dataset and verifying its efficacy using performance metrics.

2 Related Works

During the selection process, users have to consider many characteristics such as response time, security, availability, and so on, but these parameters will not be

[1] https://www.gartner.com/en/newsroom/press-releases/2021-04-21-gartner-forecasts-worldwide-public-cloud-end-user-spending-to-grow-23-percent-in-2021.

explicitly mentioned by the providers. Consideration of service-level agreement (SLA) [8] is an important parameter to ascertain the reliability of the CSP. So the users will have a hard time choosing the optimal service among different alternatives. Thus service selection is considered to be a multi-criteria decision-making (MCDM) problem. Several studies [2, 3, 10, 11] have utilized various MCDM techniques such as AHP, TOPSIS, BWM, etc., for the optimal selection of cloud services. The selection is based on the rank computed from the input given by the user. The uncertainties in the input data from the users as well as the QoS data from cloud services have been addressed with the help of fuzzy techniques [4, 9, 11]. The membership functions are defined in the fuzzy system in the form of a linguistic variable (low, medium, high) since the user will not be able to quantify the resource requirement. To avoid the error in the membership function formulation, an automated approach has been applied to tune the membership variables and construct the knowledge base. Many researchers have used ANFIS model [5] for prediction in different fields such as power plant data [12], residential load forecasting [1], Paddy production data [7], Chili price data [6], etc. The optimal value for membership function is selected based on the adaptive neural fuzzy inference system (ANFIS) approach [6]. We propose to implement ANFIS model for the prediction of cloud service provider that is applied on QWS dataset. From the previous study, we can conclude that different techniques have been applied to find optimal membership function, but none is better than the other for all kinds of applications.

3 Methodology

3.1 Adaptive Neuro Fuzzy Inference System (ANFIS)

This system combines neural network and fuzzy inference system thus combining the benefits of both the methods. It is really useful for solving nonlinear complex systems. It utilizes Sugeno fuzzy integral model to construct dynamic fuzzy inference system. Neuro fuzzy eliminates the problem in designing fuzzy models (framing IF-THEN rules) by automatically generating the rules and parameter optimization. As a result, this system utilizes both linguistic input and real-time data. [5] proposed a novel architecture and learning mechanism for fuzzy systems that utilizes neural network algorithms for the construction of a knowledge base in a fuzzy system with the appropriate membership function from the given input pairs. There are 5 layers in ANFIS model as depicted in Fig. 1, and they are:

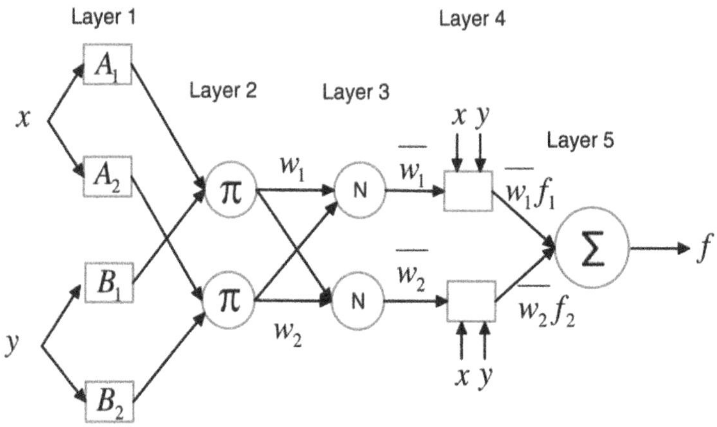

Fig. 1 ANFIS architecture

Layer 1: Every node i in this layer is a square node with a node function

$$M_i^1\mu = A_i(x), \tag{1}$$

where x is the input to node i and A_i is the linguistic label (low, med, high).
Layer 2: In this layer, the nodes are fixed, and each node output represents the firing strength of the rule

$$M_i^2\mu = W_i = \mu_{Ai}(X) * \mu_{Bi}(Y), where\ i = 1, 2. \tag{2}$$

Layer 3: The normalized firing strength of the rules obtained from the previous layer is calculated as follows:

$$M_i^3\mu = \bar{W}_i = \frac{w_i}{w_1 + w_2}, where\ i = 1, 2. \tag{3}$$

Layer 4: This layer is to defuzzify the fuzzified input, and the weighted consequent value of each rule is calculated as below:

$$M_i^4 = \bar{W}_i * f_i = \bar{W}_i(a_i * x + b_i * y + c_i), where\ i = 1, 2, \tag{4}$$

where $\bar{W}_i$ is the output of layer 3, and a_i, b_i, c_i is the parameter set. Parameters in this layer will be referred to as consequent parameters.
Layer 5: The single node in this layer computes the overall output as the summation of all incoming signals, i.e.,

$$M_i^5 = \sum_i \bar{W}_i * f_i = \frac{\sum_i \bar{W}_i * f_i}{\sum_i \bar{W}_i}. \tag{5}$$

In the fuzzy inference model, fuzzy rule is generated based on the partitioning of input data that is done by either grid partitioning or subtractive clustering. Subtractive clustering is used to generate fuzzy with a minimum number of rules. The dataset is partitioned into different clusters based on data points, and rules are framed according to the number of clusters. When the input parameters are more than 3, then grid partitioning will generate more rule base and will be less efficient.

3.2 *Tuning ANFIS Model*

The built FIS model can be tuned by changing the numbers of epochs, step size, error rate, etc., to reduce the training error. Epochs are the number of iterations for training the FIS, and more epochs mean that it would take more time to train the system. We should monitor the error rate at different epochs to identify the optimal iteration. When the epochs are increased, the training error rate would reduce, but the testing error might increase after a certain epoch. This is an indication of over-fitting of the model, and so care must be taken to avoid it. Initial step size is another parameter used in tuning the ANFIS model. In the training phase, the step size increases gradually, reaches a maximum level, and then starts to decrease. By modifying the initial step size, an ideal step size profile can be achieved. Membership function in ANFIS is obtained in two ways: i) back propagation for both input and output parameters and ii) hybrid method that consists of back propagation and least square estimation for training the input and output parameters, respectively. The training process is continued by adjusting the learning parameters until the desired RMSE (root mean square error) is reached. The model is evaluated by using the below metrics:

Mean-squared error(MSE): The average difference observed in the predicted and actual values across the whole test set.

Root mean-squared error (RMSE): It is the square root of mean-squared error. It measures the standard deviation of residuals.

Coefficient of determination (R^2): It is to determine the strength of linear relationship between two variables.

Mean absolute error (MAE): It is the difference between the measured value and the actual value.

The workflow of the proposed approach that employs ANFIS for the prediction of cloud service providers is shown in Fig. 2. The training data is clustered using weighted average clustering, and a fuzzy inference system is built. After the learning process, the system is tested using test data and validated by checking the performance metrics.

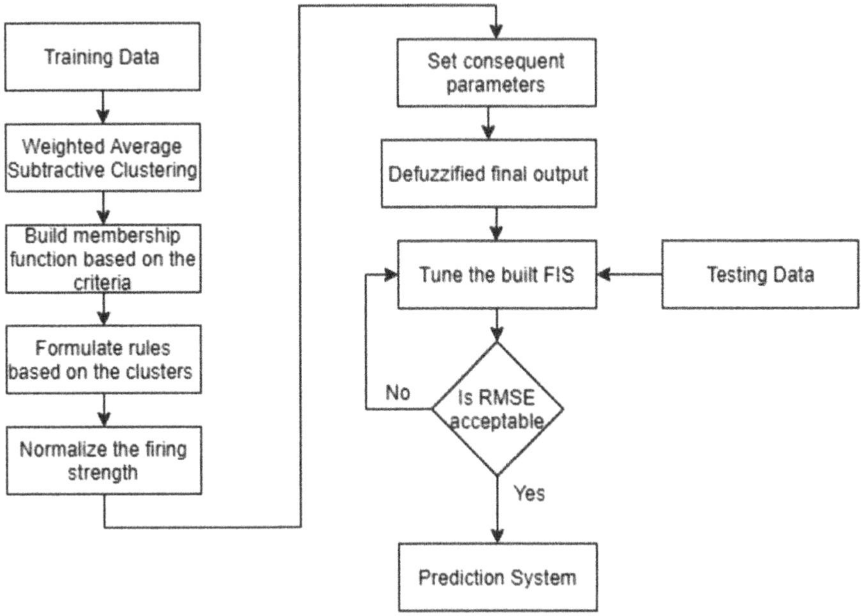

Fig. 2 Model flow

4 Results and Discussions

The proposed approach is implemented in Matlab, and the QWS[2] real-time dataset is used for the implementation. Since real-time cloud service dataset for QoS parameters along with SLA is very hard to gather, we have used web service dataset that is similar to cloud services. The attributes that are considered for the evaluation are response time, availability, throughput, successability, reliability, compliance, trust rating, and latency.

Step 1: We have filtered the web services that have trust rating less than 1 from the raw data and prepared the dataset for the learning process.

Step 2: To build fuzzy inference system (FIS), there are two types of clustering algorithms suitable for fuzzy systems, and they are fuzzy C-means (FCM) and subtractive clustering. The numbers of clusters must be specified prior for FCM, while subtractive clustering divides the data points automatically. We have shown the different clusters created by both subtractive clustering and FCM in Figs. 3 and 4, respectively. Table 1 shows the comparison of these two cluster results. We have chosen subtractive clustering for dividing the given dataset automatically. We have compared both the clustering techniques and identified that the RMSE is in acceptable range for subtractive clustering than FCM.

[2] https://github.com/wsdream/wsdream-dataset.

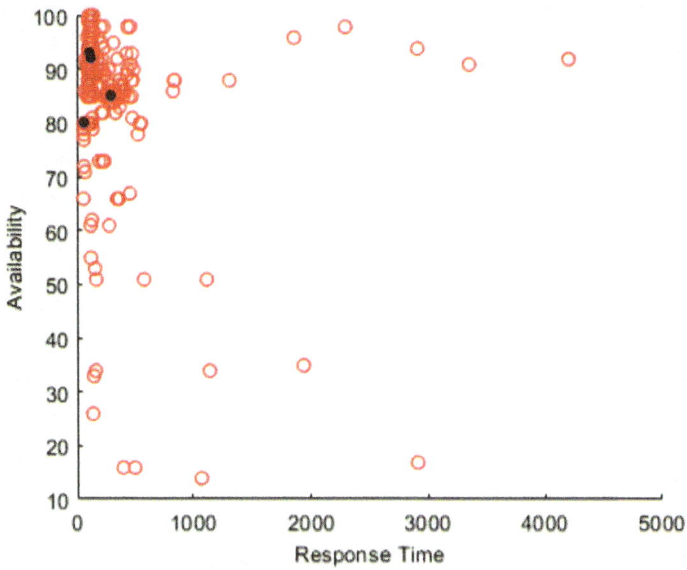

Fig. 3 Subtractive clustering

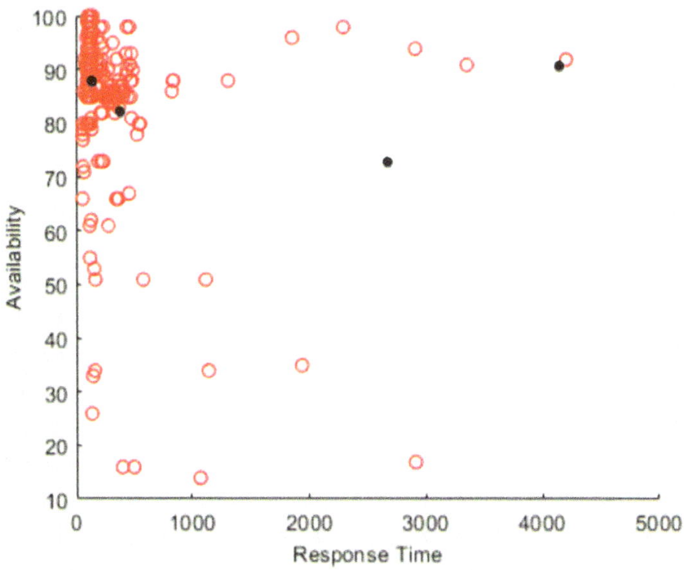

Fig. 4 FCM

Table 1 Clustering results comparison

Parameters	SC	FCM
Training RMSE	0.8884	1.7714
Testing RMSE	0.9886	1.4877
No. of nodes	101	83
No. of linear parameters	45	36
No. of nonlinear parameters	80	64
No. of parameters	125	100
No. of training data pairs	250	250
No. of checking data pairs	0	0
No. of fuzzy rules	5	4
Minimal training RMSE	0.681009	1.377767
Tuned training RMSE	0.6672	1.1593
Tuned testing RMSE	0.5101	0.9773

Table 2 Implementation parameters

Implementation technique	Subtractive clustering
Optimization technique	Hybrid optimization
No. of epochs	20,200
No. of membership functions	8
Step size	0.1,0.2,0.3,0.4,0.5

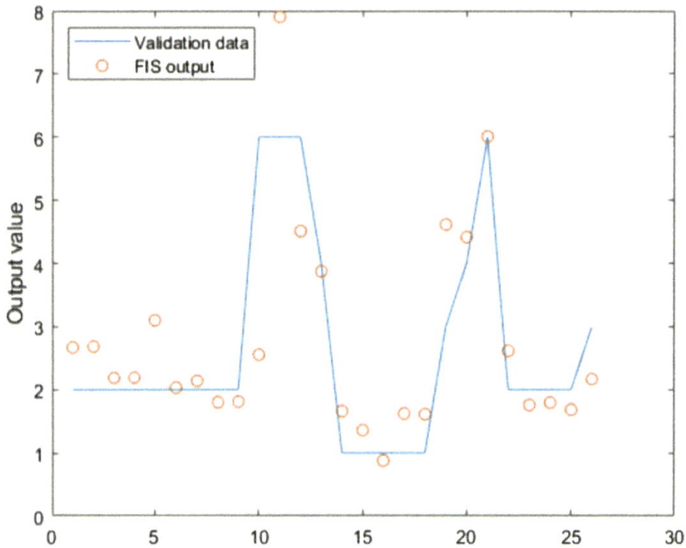

Fig. 5 Initial FIS

Step 3: Table 2 summarizes the parameters used in tuning the input membership function of the fuzzy model. This initial FIS built is shown in Fig. 5. By tuning the built FIS at different epoch levels, we have computed the RMSE, and it is shown in Table 3. We have also adjusted the initial step size and recorded the RMSE value

Table 3 RMSE at different epochs

Epochs	15	20	25	30	35	40	50
Minimal RMSE	0.7367	0.6811	0.6288	0.6021	0.5787	0.5591	0.5391

Table 4 RMSE at different step size

Step size	0.1	0.2	0.3	0.4	0.5	0.6	1
Training RMSE	0.6672	0.6529	0.6746	0.6506	0.6462	0.6267	0.4246
Testing RMSE	0.5101	0.5239	0.5202	0.5379	0.5431	0.5447	0.6182

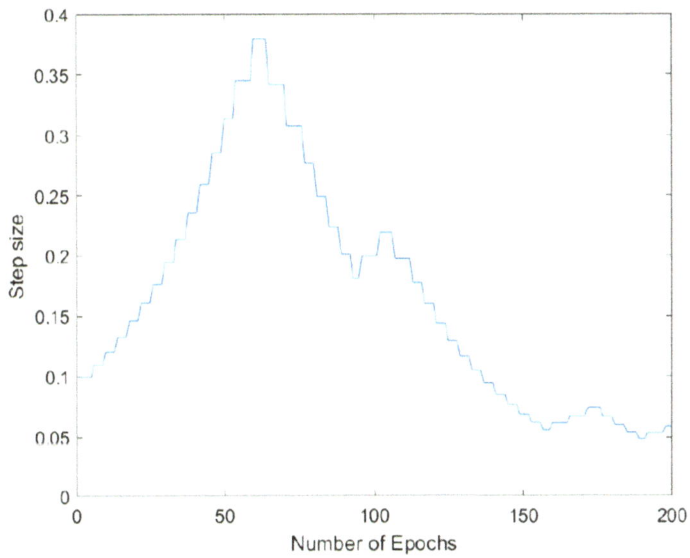

Fig. 6 Step size profile

in Table 4 to figure out the optimal step size. We have identified that the ideal step profile is achieved at 0.1, and the profile is shown in Fig. 6. The tuned FIS model with minimal RMSE is shown in Fig. 7.

Step 4: We have applied the two different optimization methods for membership function parameter training and identified that the minimal RMSE is achieved while tuning the ANFIS model with hybrid optimization technique. The built ANFIS model is shown in Fig. 8.

Step 5: To avoid over-fitting of the model, we trained the model at 200 epochs and determined the place where the over-fitting occurs. The training error reduces as the number of epochs increases and is shown in Fig. 9. But the validation error that is also called as testing error reaches a minimum at 20 epochs and then increases, which indicates the point at which over-fitting starts, and it is shown in Fig. 10.

Step 6: The model is validated by comparing the actual and predicted outputs (before and after tuning) and is shown in Fig. 11.

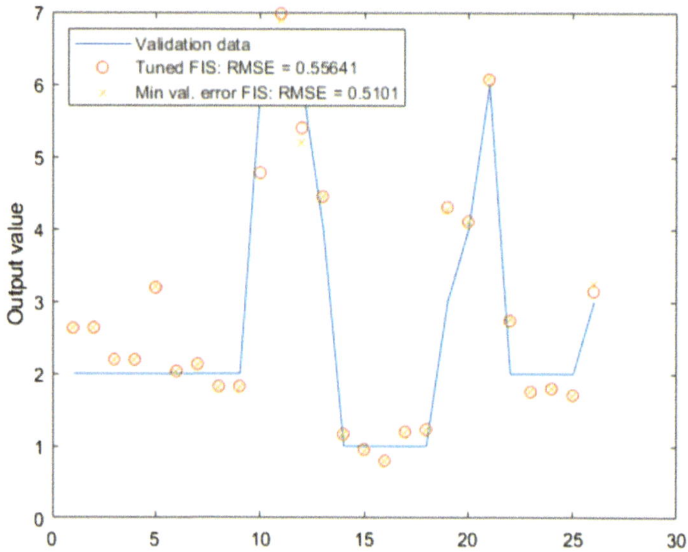

Fig. 7 Tuned FIS

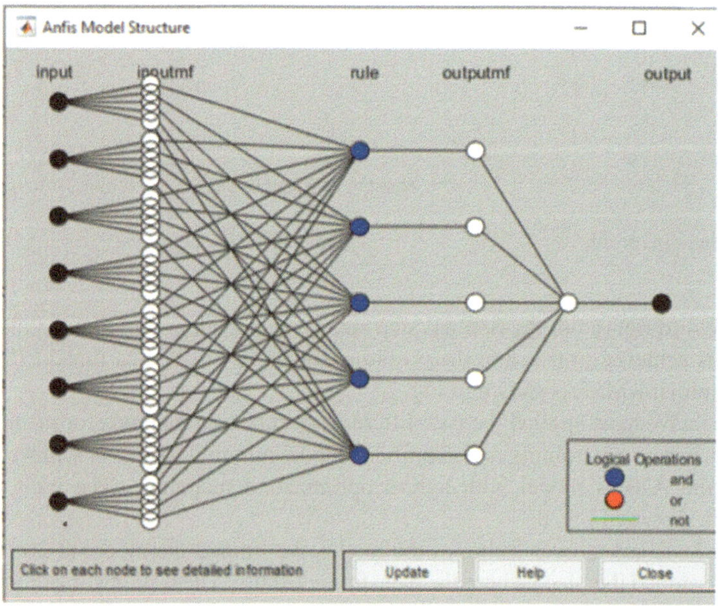

Fig. 8 ANFIS structure

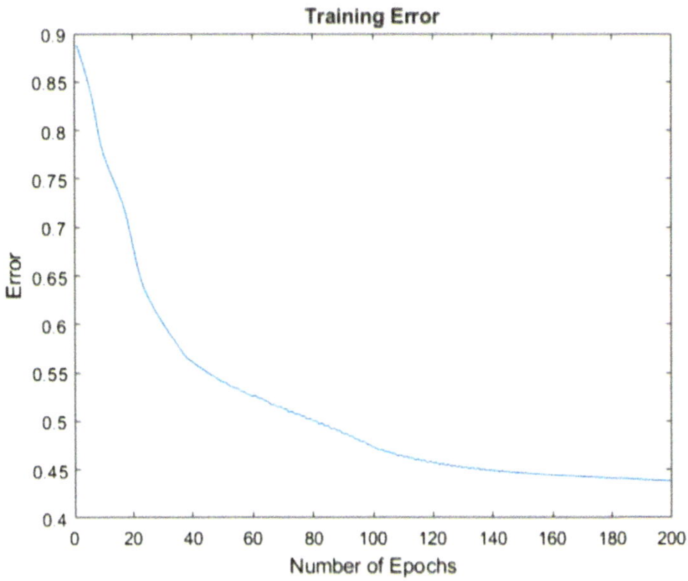

Fig. 9 Training error

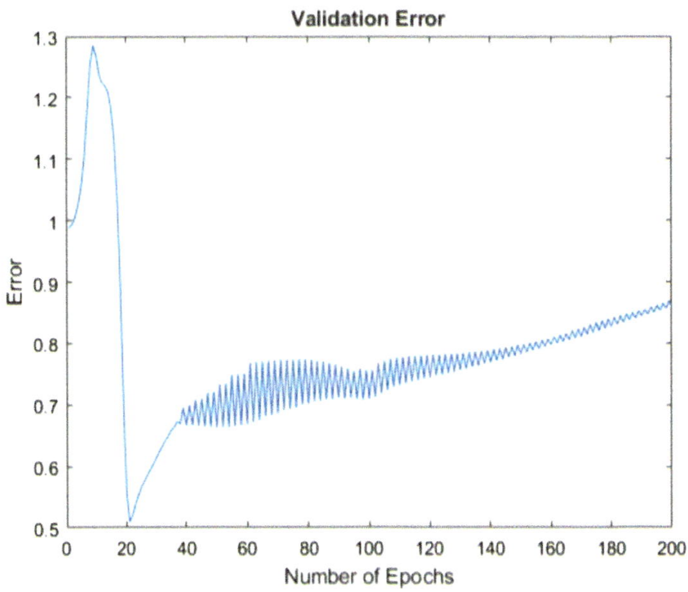

Fig. 10 Testing error

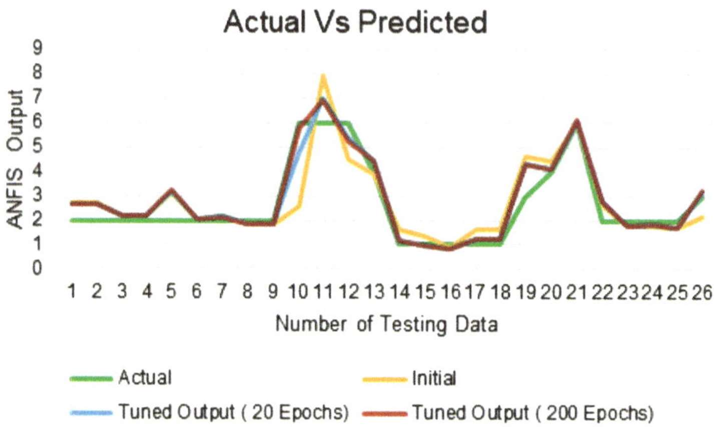

Fig. 11 Actual output vs. predicated output

Table 5 RMSE at different epochs

Measurement index	Initial	Tuned (20 epochs)	Tuned (200 epochs)
R-Squared	0.625704	0.881426	0.900338
RMSE	0.9886	0.5564	0.5101
MSE	0.97732996	0.30958096	0.26020201
MAE	0.1036884615	0.1674307692	0.1991730769
CA	0.4839728723	0.2779935738	0.2669485754

Step 7: Different performance metrics are calculated to confirm the suitability of ANFIS model for the prediction of cloud service provider, and the results are available in Table 5.

It is evident that the error rate is initially high, and after epoch 20, the tuned function gives less error rate and better R-squared value nearer to 1 to prove the prediction efficiency. Also, for the given dataset, the number of clusters is not known initially, it is better to use subtractive clustering, and also it produces lesser RMSE value than FCM. The uncertainty in the rule base is also eliminated by automatically tuning the membership function using neural network.

5 Conclusion

Many criteria have to be considered to find the optimal cloud service provider for a specific user requirement. The input parameters such as QoS attributes, workload characteristics, etc., are uncertain in nature, and so a fuzzy inference system is most suitable for this type of problem. In order to improve the accuracy of the selection, the FIS has been tuned iteratively using ANFIS for better prediction. In the proposed approach, the testing RMSE is 0.5101, and the training RMSE is 0.6672, which is in the acceptable range. The R square is 0.900338, which is nearer to 1, which proves the efficiency of the model. So we conclude that this is the suitable model for cloud service selection. This work can be extended with real-time cloud service dataset, and machine learning algorithms can be applied to improve the efficiency of the system.

References

1. S.M. Alam, M.H. Ali, A new subtractive clustering based ANFIS system for residential load forecasting, in *2020 IEEE Power & Energy Society Innovative Smart Grid Technologies Conference (ISGT)* (IEEE, Piscataway, 2020), pp. 1–5
2. S.K. Garg, S. Versteeg, R. Buyya, A framework for ranking of cloud computing services. Futur. Gener. Comput. Syst. **29**(4), 1012–1023 (2013)
3. T. Guo, R. Bahsoon, T. Chen, A. Elhabbash, F. Samreen, Y. Elkhatib, Cloud instance selection using parallel K-means and AHP, in *Proceedings of the 12th IEEE/ACM International Conference on Utility and Cloud Computing Companion* (2019), pp. 71–76
4. A. Hussain, J. Chun, M. Khan, A novel framework towards viable Cloud Service Selection as a Service (CSSaaS) under a fuzzy environment. Futur. Gener. Comput. Syst. **104**, 74–91 (2020)
5. J.S. Jang, ANFIS: adaptive-network-based fuzzy inference system. IEEE Trans. Syst. Man Cybernet. **23**(3), 665–685 (1993)
6. M. Kabir, M.M.J. Kabir, Fuzzy membership function design: An adaptive neuro-fuzzy inference system (ANFIS) based approach, in *2021 International Conference on Computer Communication and Informatics (ICCCI)* (IEEE, Piscataway, 2021), pp. 1–5
7. A. Rusgiyono, Adaptive Neuro Fuzzy Inference System (ANFIS) approach for modeling paddy production data in Central Java, in *Journal of Physics: Conference Series*, vol. 1217, No. 1 (IOP Publishing, Bristol, 2019), p. 012083
8. J. Sidhu, S. Singh, Design and comparative analysis of MCDM-based multi-dimensional trust evaluation schemes for determining trustworthiness of cloud service providers. J. Grid Comput. **15**(2), 197–218 (2017)
9. N. Tanoumand, D.Y. Ozdemir, K. Kilic, F. Ahmed, Selecting cloud computing service provider with fuzzy AHP, in *2017 IEEE International Conference on Fuzzy Systems (FUZZ-IEEE)* (IEEE, Piscataway, 2017), pp. 1–5
10. N. Upadhyay, Managing cloud service evaluation and selection. Proc. Comput. Sci. **122**, 1061–1068 (2017)
11. Z. ur Rehman, O.K. Hussain, F.K. Hussain, Iaas cloud selection using MCDM methods, in *2012 IEEE Ninth International Conference on e-business Engineering* (IEEE, Piscataway, 2012), pp. 246–251
12. C.U. Yeom, K.C. Kwak, Performance comparison of ANFIS models by input space partitioning methods. Symmetry **10**(12), 700 (2018)

Experimental Comparative Analysis on Convolutional Neural Network (CNN) and Recurrent Neural Network (RNN) on Aspect-Level Sentiment Analysis

Theresa V. Cherian, Getzi Jeba Leelipushpam Paulraj, Immanuel Johnraja Jebadurai, and Jebaveerasingh Jebadurai

Abstract Sentiment analysis is a process of inferencing subjective text with sentiments. Huge data generated by social media, business, medical industry, e-commerce, engineering, and academia are analyzed to understand customer satisfaction, trends, experience, opinion, and influence. Various algorithms have been proposed in the literature to analyze the generated data. This paper summarizes various types of sentiment analysis and the process of sentiment analysis and discusses the application of two main deep learning architectures convolutional neural networks (CNN) and recurrent neural networks (RNN) in sentiment analysis. The experimental analysis also has been carried out on these two architecture and performance metrics, viz., accuracy, loss, precision, and recall, which have been compared.

Keywords Sentiment classification · CNN · RNN · RecNN · Machine learning · Deep learning · NLP · Opinion mining · Data retrieval · Speech translation · Sentiment analysis · Long short-term memory (LSTM)

1 Introduction

Natural language processing (NLP) aims at providing computers with human-like features to understand text and words using artificial intelligence (AI) tools. The various tasks of NLP include speech recognition, tagging, data analysis, reference resolution, and language generation. Data analysis in NLP is performed in various ways [1].

T. V. Cherian · G. J. L. Paulraj (✉) · I. J. Jebadurai · J. Jebadurai
Karunya Institute of Technology and Sciences, Coimbatore, India
e-mail: getzi@karunya.edu; jebaveerasingh@karunya.edu

A. Haldorai et al. (eds.), *4th EAI International Conference on Big Data Innovation for Sustainable Cognitive Computing*, EAI/Springer Innovations in Communication and Computing, https://doi.org/10.1007/978-3-031-07654-1_2

Topic analysis: Topic analysis (also called topic detection, topic modeling, or topic extraction) is a technique that organizes and understands large collections of text data, by assigning "tags" or categories according to each text's topic or theme.

- Time Series Analysis: Time series analysis is a statistical method to analyze the past data within a given duration of time to forecast the future. It comprises an ordered sequence of data at an equally spaced interval.
- Sentiment Analysis: Sentiment analysis (also known as opinion mining or emotion AI) refers to the use of natural language processing, text analysis, computational linguistics, and biometrics to systematically identify, extract, quantify, and study affective states and subjective information.
- Network Analysis: Analysis performed using different network algorithms such as Bayesian networks, neural trees, and radial basis function (RBF) networks.

Sentiment analysis is a widely used technique applied to customer reviews, survey responses, and user reactions in online and social media which improves business service to clinical medicine [11, 12].

1.1 Sentiment Analysis

Sentiment analysis aims at retrieving useful insights from junk unstructured data. It helps the business to grow by analyzing various data that include online product reviews, movie reviews, airline reviews, education reviews, and social media reviews [6].

Various machine learning techniques have been traditionally used to analyze sentiment data [12, 13]. However, it is challenging to process these data using a machine learning algorithm due to the following reasons. Firstly, the data is highly unstructured. Even though machine learning algorithms deal with unstructured data, streamlining raw data for processing is a cumbersome job using machine learning algorithms. Secondly, the amount of data generated is huge, which expects more iteration of technique with those data to yield better accuracy. Thirdly, feature extraction is vital for any classification problem which is dealt with as a separate module in machine learning techniques. Various steps involved in sentiment analysis are as follows [2, 3]:

1. Data Collection: During data collection, data is fetched from blogs, websites, social media, etc., and using application programming interfaces. These fetched data are processed using natural language processing (NLP) tools and moved forward to the pre-processing stage.
2. Data Pre-Processing: During data pre-processing stage, stop words, punctuations and special characters are removed from the extracted data. Proper pre-processing improves the efficiency of the analysis system.
3. Sentiment Detection: The pre-processed data is converted into sentences and categorized into a subjective and objective category. The objective expression has to be removed for accurate classification of the text.

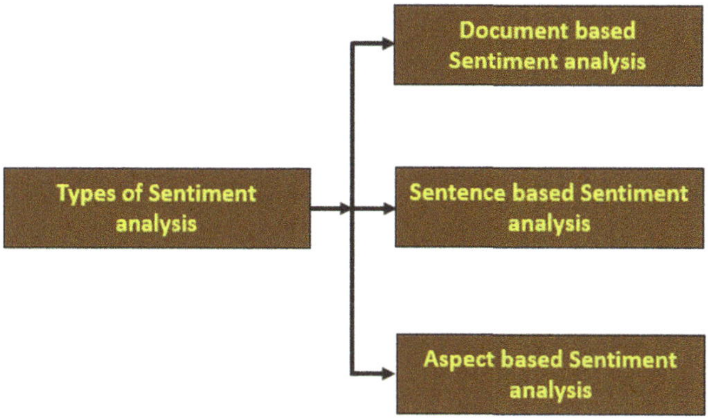

Fig. 1 Types of sentiment analysis

Fig. 2 Example of aspect-based sentiment classification [4]

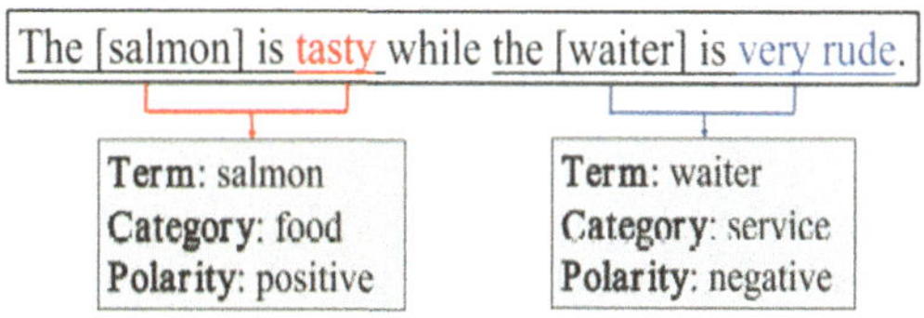

4. Sentiment Classification: The subjective expression is processed using suitable machine learning or deep learning algorithms for classifying as positive, negative, good, bad, like, and dislike.
5. Output Preparation: The performance of the applied algorithms is evaluated and pruned for better performance. The outputs are also visualized in graphs and charts using data visualization tools.

There are various types of sentiment analysis, viz. [7, 11], document-based sentiment analysis, sentence-based sentiment analysis, and aspect-based sentiment analysis. The types are depicted in Fig. 1. In document-based sentiment analysis, the whole document is considered for analysis. The document as a whole is classified as positive or negative. In sentence-based analysis, every sentence is extracted and analyzed to classify as positive or negative. However, these two analyses do not provide accurate classification and depicted in the example shown in Fig. 2.

The example quotes, "The Salmon is tasty while the waiter is very rude." This sentence has two aspects viz., food and service. The subject has a positive opinion of the food but has a negative opinion of the service. These finer details can be classified using aspect-based sentiment analysis. This involves the extraction of aspects from a sentence, opinion identification, and aspect-level classification. Various deep learning architectures have been applied for classification [11]. This paper explores an experimental comparative study of convolutional Neural Networks (CNN) and recurrent neural networks (RNN) on aspect-based sentiment analysis.

The remainder of this paper is organized as follows: Section 2 explains the CNN and RNN architecture in classifying sentiments. Section 3 explains the dataset used for experimental comparative analysis. Section 5 displays the results and discussion about the comparison. The next section concludes with the future scope of the paper.

2 CNN and RNN in Sentiment Analysis

Various deep learning architectures [9, 10] have been proposed to analyze the sentiments. As CNN and RNN form the basic deep learning architectures, the performance of those architectures is analyzed on three different tweet datasets. This section explains the CNN and RNN architecture in detail.

2.1 Convolutional Neural Network-Based Analysis

Convolutional neural networks have steps, viz., text pre-processing, vectorization, defining the model, training, and testing the model. The CNN-based sentiment analysis is depicted in Fig. 3. The initial step for any sentiment analysis is data pre-processing. During the pre-processing step, the sentences in the documents are split into words. This step is called tokenization and is performed using the NLTK tokenizer. Now the document will contain symbols, digital, and punctuation which do not provide any useful insight for our classification. These symbols, digits, and punctuations are removed in the next step. The next step of pre-processing involves the removal of stop words from the documents. This is performed by downloading the list of stop words from the NLTK library. The final step is lemmatization. This is the process of capturing the context of a word and converting it into its base form. This is a very important step for text classification.

To extract information from the text using deep learning models [14, 15], the words need to be converted into numbers. This process of converting text into real numbers is called vectorization. Various techniques, viz., bag-of-words, term frequency-inverse document frequency (TF-IDF), and Google Word2vec, are available for vectorization. In this study, we have used TF-IDF vectorization.

The convolutional neural deep learning model has embedding, convolution, Global max pooling, dropout, dense, and again a dropout layers [5]. The convolutional neural network model is shown in Fig. 4.

After defining the model, it is trained and tested with datasets, and results are obtained.

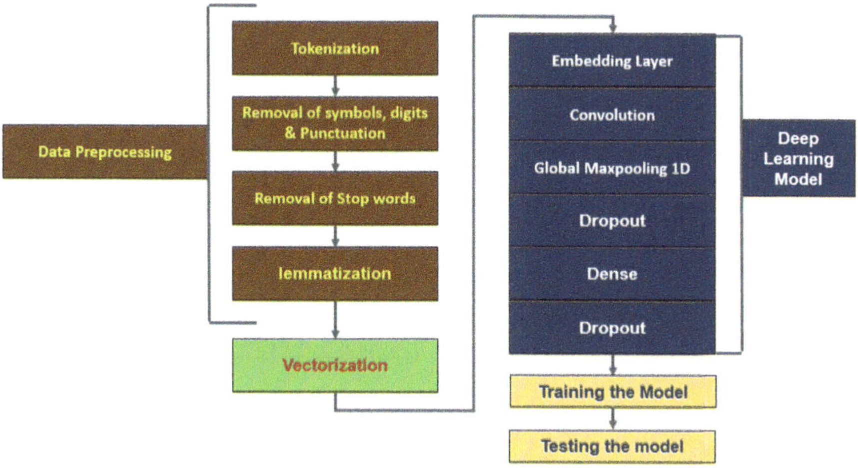

Fig. 3 Step-by-step sentiment analysis using CNN architecture

```
Model: "sequential_2"
_________________________________________________________________
Layer (type)                 Output Shape              Param #
=================================================================
embedding_2 (Embedding)      (None, 176, 100)          1243000
_________________________________________________________________
conv1d_1 (Conv1D)            (None, 174, 1024)         308224
_________________________________________________________________
global_max_pooling1d_1 (Glob (None, 1024)              0
_________________________________________________________________
dropout_8 (Dropout)          (None, 1024)              0
_________________________________________________________________
batch_normalization_4 (Batch (None, 1024)              4096
_________________________________________________________________
dropout_9 (Dropout)          (None, 1024)              0
_________________________________________________________________
dense_4 (Dense)              (None, 2048)              2099200
_________________________________________________________________
dropout_10 (Dropout)         (None, 2048)              0
_________________________________________________________________
batch_normalization_5 (Batch (None, 2048)              8192
_________________________________________________________________
dropout_11 (Dropout)         (None, 2048)              0
_________________________________________________________________
dense_5 (Dense)              (None, 3)                 6147
=================================================================
Total params: 3,668,859
Trainable params: 3,662,715
Non-trainable params: 6,144
```

Fig. 4 Convolutional neural network model

```
Model: "sequential"
_________________________________________________________________
Layer (type)                 Output Shape              Param #
=================================================================
embedding (Embedding)        (None, 176, 100)          1243000
_________________________________________________________________
lstm (LSTM)                  (None, 256)               365568
_________________________________________________________________
dropout (Dropout)            (None, 256)               0
_________________________________________________________________
batch_normalization (BatchNo (None, 256)               1024
_________________________________________________________________
dropout_1 (Dropout)          (None, 256)               0
_________________________________________________________________
dense (Dense)                (None, 512)               131584
_________________________________________________________________
dropout_2 (Dropout)          (None, 512)               0
_________________________________________________________________
batch_normalization_1 (Batch (None, 512)               2048
_________________________________________________________________
dropout_3 (Dropout)          (None, 512)               0
_________________________________________________________________
dense_1 (Dense)              (None, 3)                 1539
=================================================================
Total params: 1,744,763
Trainable params: 1,743,227
Non-trainable params: 1,536
_________________________________________________________________
```

Fig. 5 Recurrent neural network model

2.2 *Recurrent Neural Network (RNN)-Based Analysis*

Recurrent neural network (RNN) is a deep learning model that evaluates time series data. Interestingly the RNN holds a memory for storing previous outputs, and the present computation depends on stored previous outputs. The pre-processing steps are similar to that of convolutional neural networks. The RNN model used for analysis is depicted in Fig. 5.

3 Performance Metrics

The improved classification techniques using the feature selection technique are evaluated against the following performance metrics [8]:

3.1 Accuracy

The accuracy is the ratio between the number of correctly classified instances to that of a total number of instances. The formula for accuracy is given in Eq. (1).

$$\text{Accuracy} = \frac{\text{TP} + \text{TN}}{\text{TP} + \text{FP} + \text{FN} + \text{TN}} \quad (1)$$

The terminologies associated with the metric are:

True positive (TP): True positives are cases when actual true cases are correctly predicted as true.
True negative (TN): True negatives are cases when actual false cases are correctly predicted as false.
False positive (FP): False positives are cases when actual false cases are wrongly predicted as true.
False negative (FN): False negatives are cases when actual true cases are wrongly predicted as false.

4 Dataset

The evaluation was carried out on US Airline sentiment analysis and presidential debates dataset. The dataset is collected by scraping Twitter from Feb 2015 from the tweets of passengers who have travelled through major airlines of the USA [16]. The dataset has 15 columns and 14,640 rows. The second dataset used was the presidential dataset that was accumulated from the tweets during US elections in 2020. The transcripts of the videos are collected in the form of a dataset. The datasets are analyzed using CNN and RNN deep neural network architectures, and performance metrics, viz., accuracy and loss function, are measured. The performance analysis is shown in the following section.

5 Performance Analysis on CNN and RNN Architectures

The model accuracy of CNN and RNN for the airline sentiment dataset is shown in Figs. 6 and 7.

Figure 6 shows the accuracy and loss of CNN architecture for the airline datasets. From the figure, it is understood that the accuracy for the training phase is high. However, when testing the architecture using the test set, it is understood that the accuracy has reached an average of 75%. Similarly, the loss is less for the training phase, and it elevates up to 0.7 when the testing dataset is applied. Figure 7 shows the performance analysis of RNN for the airline dataset. From the figure it is understood that the testing accuracy of the RNN architecture for the testing phase is 78%;

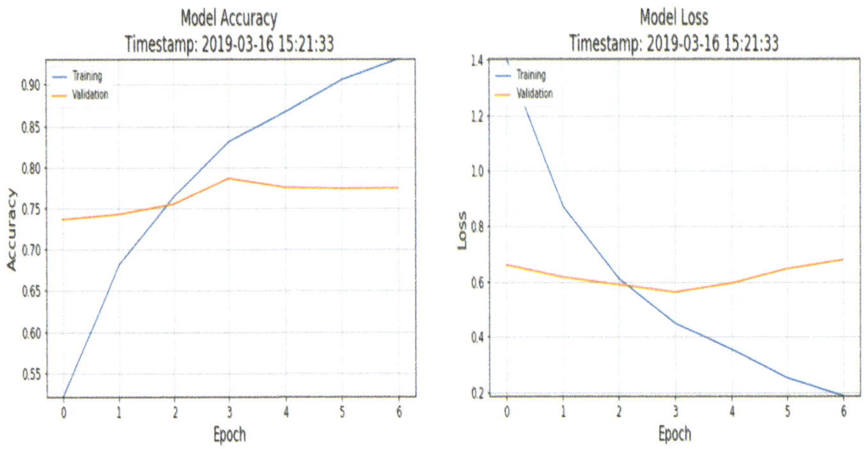

Fig. 6 Accuracy and loss of CNN

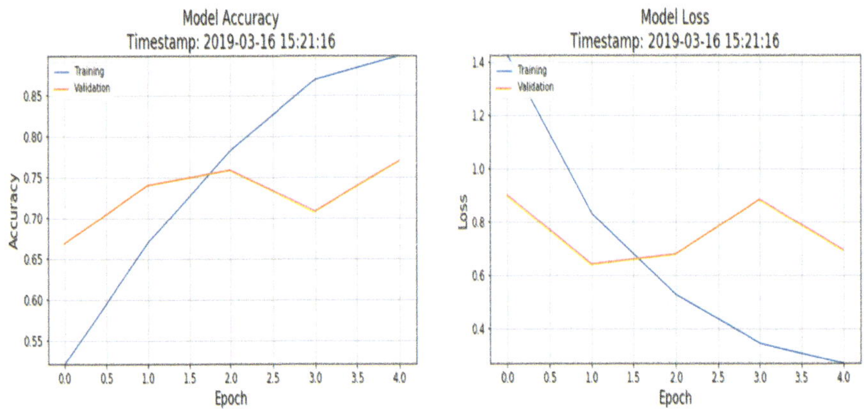

Fig. 7 Accuracy and loss of RNN

however, the loss function is around 0.87. Model performance is also evaluated for CNN and RNN, and the score for CNN analyzed using the airline sentiment dataset is 79.64 and for RNN architecture is 74.27. The performance analysis of CNN and RNN architectures using the presidential debate dataset is shown in Figs. 8 and 9.

Figure 8 shows the accuracy and loss of CNN architecture. It is understood that the accuracy of CNN during the testing dataset is 65% and the loss is also abnormally high. Similarly, the performance of the RNN architecture is shown in Fig. 9. Accuracy is around 65%, and testing loss is high compared to the training loss. Model performances of CNN and RNN are calculated as 77.77 and 75.28, respectively. The CNN architecture has a better performance compared with RNN architecture when tested with both the US airline dataset and the presidential dataset.

Performance analysis of RNN and CNN for both datasets is tabulated in Table 1.

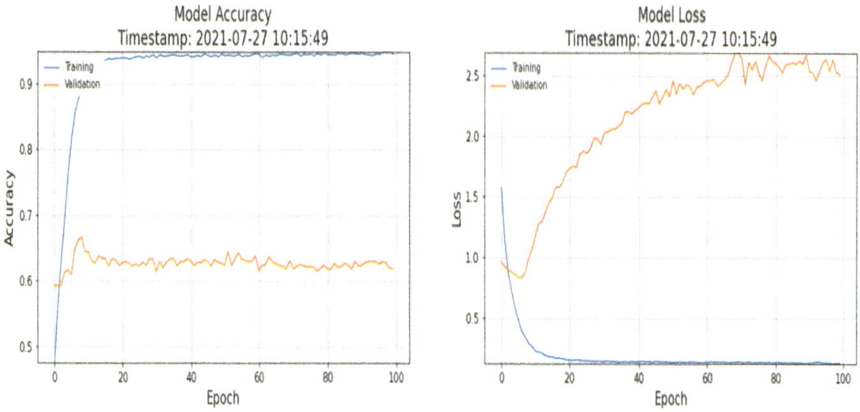

Fig. 8 Accuracy and loss for CNN

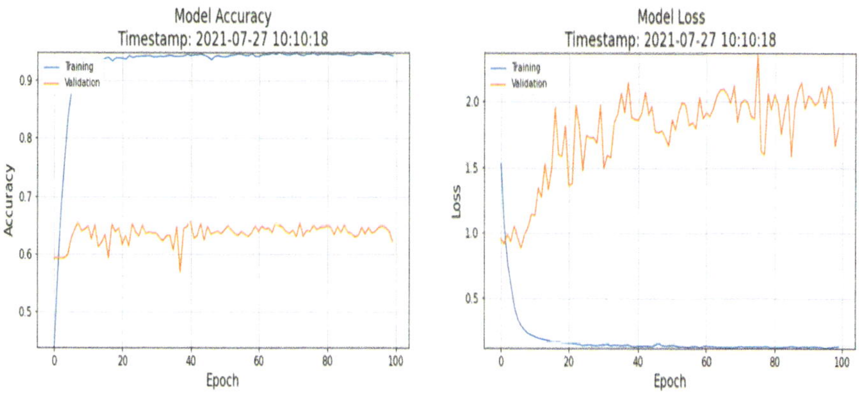

Fig. 9 Accuracy and loss for RNN

Table 1 Performance analysis on RNN and CNN on both datasets

Dataset	Model	Accuracy (%)	Loss	Performance
US airline dataset	RNN	78	0.87	74.27
	CNN	75	0.70	79.64
Presidential dataset	RNN	65	High	75.28
	CNN	65	High	77.77

6 Future Research Directions

By evaluating the performance of the CNN and RNN architecture on sentiment analysis, it is understood that the accuracy and loss function is not appreciably good for the analysis. Hence the below future research directions have been proposed:

- A hybrid of CNN and RNN architecture can be proposed.
- Attention mechanisms to capture spatial and temporal parameters can be incorporated to improve the accuracy and reduce the loss of the analysis.
- Deeper neural architecture with more epochs can be implemented.
- Weights have to be regularized.

In the future, this paper will be enhanced with a novel deep neural network model for aspect-level sentiment analysis.

7 Conclusion

Sentiment analysis improves the business in terms of prediction, diagnosis, and recommendations. Deep learning architecture analyses chuck of data and aid sentiment analysis. Among the deep learning architecture, CNN and RNN form the basic architectures for sentiment analysis. This paper performs an experimental analysis on sentiment analysis using CNN and RNN architecture on US airline and presidential debate databases. The experimental results prove that the CNN architecture provides a better score compared with the RNN architecture. However, to improve the accuracy of the analysis, future research directions are also discussed in this paper.

References

1. H. Liu, I. Chatterjee, M. Zhou, X.S. Lu, A. Abusorrah, Aspect-based sentiment analysis: A survey of deep learning methods, in *IEEE Transactions on Computational Social Systems*, vol. 7, Issue, 6, (Institute of Electrical and Electronics Engineers Inc., 2020), pp. 1358–137. https://doi.org/10.1109/TCSS.2020.3033302
2. A. D'Andrea, F. Ferri, P. Grifoni, T. Guzzo, Approaches, tools and applications for sentiment analysis implementation. Int. J. Comput. Appl. **125**(3), 26–33 (2015). https://doi.org/10.5120/ijca2015905866
3. J. Jie Zhou, J.X. Huang, Q. Chen, Q.V. Hu, T. Wang, L. He, Deep learning for aspect-level sentiment classification: Survey, vision, and challenges, in *IEEE Access*, vol. 7, (Institute of Electrical and Electronics Engineers Inc., 2019), pp. 78454–78483. https://doi.org/10.1109/ACCESS.2019.2920075
4. S.C. Nistor, M. Moca, D. Moldovan, D.B. Oprean, R.L. Nistor, Building a twitter sentiment analysis system with recurrent neural networks. Sensors **21**(7) (2021). https://doi.org/10.3390/s21072266

5. W. Yin, K. Kann, M. Yu, H. Schütze, Comparative study of CNN and RNN for natural language processing, (2017), http://arxiv.org/abs/1702.01923
6. N.C. Dang, M.N. Moreno-García, F. De la Prieta, Sentiment analysis based on deep learning: A comparative study. Electronics (Switzerland) **9**(3) (2020). https://doi.org/10.3390/electronics9030483
7. S. Das, A.K. Kolya, Predicting the pandemic: Sentiment evaluation and predictive analysis from large-scale tweets on Covid-19 by deep convolutional neural network. Evol. Intel. (2021). https://doi.org/10.1007/s12065-021-00598-7
8. S. Pathak, Twitter sentiment analysis using different algorithms. Int. J. Res. Appl. Sci. Eng. Technol. **8**(9), 1023–1026 (2020) https://doi.org/10.22214/ijraset.2020.31647
9. G.H. Neha, S. Pande, A. Khamparia, V. Bhagat, N. Karale, Twitter sentiment analysis using deep learning. IOP Conf. Ser.: Mater. Sci. Eng. **1022**(1) (2021). https://doi.org/10.1088/1757-899X/1022/1/012114
10. S. Kamiş, D. Goularas, Evaluation of deep learning techniques in sentiment analysis from twitter data, in *Proceedings – 2019 International Conference on Deep Learning and Machine Learning in Emerging Applications, Deep-ML 2019*, June 2019, 12–17 (2019), https://doi.org/10.1109/Deep-ML.2019.00011
11. S. Yi, X. Liu, Machine learning based customer sentiment analysis for recommending shoppers, shops based on customers' review. Complex Intell. Syst. **6**(3), 621–634 (2020). https://doi.org/10.1007/s40747-020-00155-2
12. A. Pak, P. Paroubek, Twitter as a corpus for sentiment analysis and opinion mining. (n.d.), http://tumblr.com
13. M.M. Agüero-Torales, J.I. Abreu Salas, A.G. López-Herrera, Deep learning and multilingual sentiment analysis on social media data: An overview. Appl. Soft Comput. **107** (2021). https://doi.org/10.1016/j.asoc.2021.107373
14. L.C. Chen, C.M. Lee, M.Y. Chen, Exploration of social media for sentiment analysis using deep learning. Soft. Comput. **24**(11) (2020). https://doi.org/10.1007/s00500-019-04402-8
15. A. Yadav, D.K. Vishwakarma, Sentiment analysis using deep learning architectures: A review. Artif. Intell. Rev. **53**(6) (2020). https://doi.org/10.1007/s10462-019-09794-5
16. H. Saif, Y. He, H. Alani, Semantic sentiment analysis of twitter (2012), http://iswc2012.semanticweb.org/

AI-Enabled Pregnancy Risk Monitoring and Prediction: A Review

Vidyalekshmi Chandrika and Simi Surendran

Abstract Accessibility to antenatal healthcare services remains a critical challenge in rural India. According to a recent survey, the mortality ratio of pregnant mothers is 113 for each 100,000 live births from 2016 to 2018 period. Proper follow-ups and patient-specific lifestyle changes are required to get rid of many complications in pregnancy. Due to the lack of qualified professionals, non-uniform accessibility of facilities, and affordability of expenses, the majority of people are not receiving proper medical care. AI-enabled remote monitoring systems improve patient-centred access to quality services and guidance for proactive self-management. In this chapter, we review various AI and machine learning techniques applied for remote pregnancy monitoring and risk prediction.

Keywords Machine learning · Pregnancy risk prediction · Remote health monitoring

1 Introduction

Pregnancy is a physiological process of developing a foetus inside a female body, a period from conception to birth. A pregnancy in which complications are likely to occur in the case of maternal or foetal health during gestation or childbirth is differentiated as a high-risk pregnancy by clinicians. These risks may be due to several reasons such as gestational diabetes or preeclampsia. In India, around 20–30% of pregnancies are classified as high risk, accounting for 75% of perinatal morbidity and death. Adequate nutrient intake is a vital factor that aids in the healthy outcome of pregnancy. According to a survey [40], almost 25% of reproductive-age

V. Chandrika (✉) · S. Surendran
Department of Computer Science and Engineering, Amrita Vishwa Vidyapeetham, Amritapuri, India
e-mail: simisurendran@am.amrita.edu

A. Haldorai et al. (eds.), *4th EAI International Conference on Big Data Innovation for Sustainable Cognitive Computing*, EAI/Springer Innovations in Communication and Computing, https://doi.org/10.1007/978-3-031-07654-1_3

women in India have a body mass index of less than 18.5 kg/m, indicating that they are malnourished.

Prenatal care may assist women in avoiding difficulties and educating them ensures a healthy pregnancy. Women who receive regular prenatal care are less likely to have pregnancy complications. Following a healthy and balanced diet, exercising regularly as prescribed by your doctor, and avoiding contact with harmful chemicals/radiation can help reduce the risk of pregnancy. Pre-existing disorders such as high blood pressure and diabetes must be managed carefully to avoid serious complications and their consequences. Obstetric caregivers had little knowledge about the events unfolding in the interior of the gravid uterus before the discovery of x-rays and their introduction to the medical field. The development and advancement of medical ultrasonography resulted in the spectacular growth of its applications in obstetric imaging [5].

Continuous monitoring of maternal parameters can significantly reduce the risk of unfavourable foetal outcomes. There are several studies dealing with continuous health monitoring with the help of the Internet of Things (IoT) [2, 43, 44, 48]. Recently, machine learning (ML) and IoT techniques are widely applied for remote monitoring of pregnant women to reduce the maternal risks and morbidity rate. Machine learning applications prove to be efficient in the analysis of data and identifying hidden patterns. These techniques further help in predicting the adverse outcomes by learning from a vast range of varied databases. ML techniques coupled with predictive analysis can render effective personalized healthcare or suggestions on behavioural modifications.

This chapter presents a review of machine learning techniques applied for pregnancy monitoring, challenges in continuous monitoring systems, sources of datasets, and future research directions. This chapter is organized as follows. Section 2 introduces the background and terminologies. Section 3 provides an extensive review of supervised and unsupervised machine learning algorithms applied in pregnancy monitoring. Section 4 discusses the open research challenges and future directions.

2 Pregnancy Risks

2.1 Preterm Labour

Labour that occurs before the 37th week of pregnancy is referred as premature labour. Any baby born before 37 weeks is more likely to have health problems because organs such as the liver, brain, and lungs complete their development between 39 and 40 weeks before full birth. Infections, a shorter cervix, and previous premature births all increase the risk of preterm labour [19].

2.2 Preeclampsia

High blood pressure usually occurs when the arteries that carry blood from the heart to other parts of the body constrict, which is called hypertension. Increased arterial pressure makes it more difficult for blood to enter the placenta, which gives nutrients and oxygen to the foetus. Decreased blood flow can inhibit foetal growth and increase a woman's risk of premature birth. Preeclampsia is caused by high blood pressure, which occurs after 20 weeks of gestation [31, 46].

2.3 Gestational Diabetes Mellitus

A type of glucose intolerance that develops or is identified during pregnancy [36] is termed gestational diabetes mellitus. Hormonal changes during pregnancy cause the body to either not create enough insulin or not utilize it effectively, resulting in gestational diabetes. Rather, glucose accumulates in your bloodstream, leading to diabetes. Controlling gestational diabetes by following a treatment plan prescribed by a health care provider is one of the best ways to prevent problems related to high blood sugar during pregnancy [18].

2.4 Ectopic Pregnancy

Ectopic pregnancy usually occurs when the fertilized egg is placed outside the uterus, usually in the fallopian tube. It turns out to be a life-threatening condition if not treated on time.

2.5 Infections

Sexually transmitted infections can occur during pregnancy or childbirth, causing complications for both the pregnant woman and the baby. Some infections are transmitted from mother to baby through the reproductive tract during childbirth, while others can infect the foetus at any time throughout the pregnancy period [10].

2.6 Depression Anxieties

According to the American College of Obstetricians and Gynecologists, 14% to 23% of pregnant women suffer from depression [10]. Depression during pregnancy can impair a person's ability to take care of themselves. This can later lead to malnutrition, alcohol abuse, and inadequate antenatal attendance. All of these factors endanger a woman's physical and mental health, causing problems with the growth and development of the foetus [47].

2.7 Other Complications

Other causes of pregnancy complications include iron deficiency or anaemia, persistent nausea, and vomiting.

3 AI Techniques Applied in Maternal Healthcare

This section discusses the various machine learning methods adopted to monitor the gestation period and predict the complications likely to arise during this period. Figure 1 shows the summary of various AI techniques applied to predict pregnancy risks. Most of these researches utilized the electrohysterography (EHG) and cardiotocography (CTG) datasets to predict the preterm birth [15, 17, 25, 53]. Based on a comprehensive review of the literature, we observed that the majority of research in this domain focuses on the classification of foetal health and the prediction of preterm birth. Supervised learning, unsupervised learning, neural networks, genetic algorithms, and deep learning techniques were widely used to monitor and predict maternal risks.

3.1 Supervised Learning

Supervised machine learning uses labelled dataset to train algorithms that classify data or predict outcomes. Several supervised learning approaches such as logistic regression, k-nearest neighbour, support vector machine, XGboost, and decision trees were proposed for diagnosis and prediction related to pregnancy care. The major risk factors predicted using these techniques are foetal health status, preterm delivery, miscarriage, ectopic pregnancy, preeclampsia, impact of dietary intake, and relation between exposure to particulate matter$_{10}$ with congenital heart disease.

3.1.1 Foetal Health Risk Prediction

A number of supervised learning approaches were applied to predict the risk associated with pregnancy. For example, in [17], the authors proposed a three-step classification approach to classify the foetal health into normal, suspicious, and

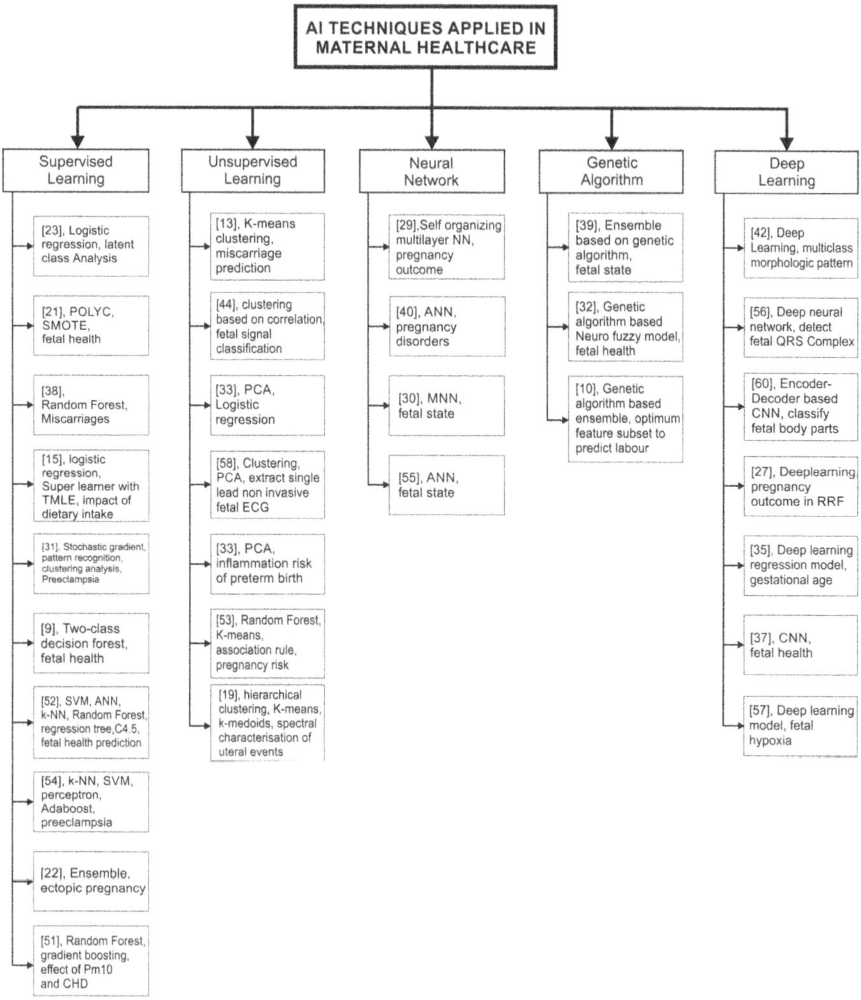

Fig. 1 AI techniques applied in maternal healthcare

pathological based on the cardiotocographic dataset [9]. In the first phase, latent class analysis (LCA) was utilized to estimate posterior probabilities of individual records, and later maximum posterior probability is used to determine the foetal health status. The LCA method was combined with a logistic regression approach in two different schemes, an ordinal classification and a multi-class classification scheme.

A mobile application system that monitors and performs prognosis of remote patients seeking clinical treatments was proposed in a work [1]. A clinical dataset of 96 pregnant women was trained using two-class decision forest algorithm, and the foetal health status is predicted with an accuracy of 89.5%. When the model applied to the real-life testing with 16 users, accuracy reported was 87.5%.

3.1.2 Preterm Delivery

For the prediction of preterm delivery, the authors [15] considered different classifiers that included linear discriminant classifier, quadratic discriminant classifier, uncorrelated normal density-based classifier, polynomial classifier, logistic classifier, k-NN, decision tree, Parzen classifier, and the support vector classifier. For the study, they utilized an open-source PhysioNet term–preterm EHG dataset [54] containing 300 records that includes 38 preterm and 262 term deliveries. The minority preterm class was oversampled by using the synthetic minority oversampling technique [7], whereas different cross-validation methods were explored to evaluate the dataset against other similar studies. Using the polynomial classifier, the technique outperformed with 96% sensitivity, 90% specificity, and 95% area under the curve.

A comparative study [50] on the performance of eight different machine learning algorithms on the UCI CTG dataset [8] was one among the reviewed works in preterm birth prediction. The authors compared artificial neural network, support vector machine, k-NN, random forest, classification and regression trees, logistic regression, C4.5, and radial bias function network. The goal of these machine learning approaches was to categorize the foetal health into pathological or normal classes utilizing features of data received from uterine contraction and foetal heart rate signals. The dataset comprises 1831 instances with 21 features and obtained a maximum accuracy of 99.2%.

3.1.3 Miscarriage Prediction

One of the most extensively utilized multi-functional imaging techniques for monitoring and identifying early pregnancy events is ultrasound imaging. The gestational sac that is the earliest symptom and quantifiable aspect of early pregnancy is assessed manually using ultrasound imaging [23]. In [33], the authors used six ML algorithms to predict pregnancy loss risks after in vitro fertilization-embryo transfer. They analyzed 66,706 records of data collected from transvaginal scans during the first trimester of pregnancy and achieved high predictive accuracy for the random forest classifier.

3.1.4 Preeclampsia Prediction

In [26], the authors aimed to predict the occurrence of preeclampsia after 34 weeks of pregnancy using different machine learning models such as stochastic gradient descent, logistic regression, decision tree model, naive bayes classification, support vector machine, and random forest algorithm. The stochastic gradient boosting model was confirmed to give the best prediction performance with an accuracy of 0.973 and a false positive rate of 0.009 on the electronic medical record of 11,006 pregnant women who received prenatal care at Yonsei University Hospital. The parameters included in the prediction models were chosen using pattern recognition and cluster analysis. In another research [56], a comparative study of logistic regression, K-nearest neighbour, decision tree, random forest, multilayer perceptron neural network, support vector machines, and naive Bayes classifier is performed to compare the preeclampsia prediction. Using a dataset of 352 pregnant women with 17 clinical features from Kenya, the random forest classifier performed better than other models. A mobile application for predicting preeclampsia and a recommendation system for high risk of preeclampsia patients has been suggested in [52]. For prediction, the authors used a soft voting ensemble learning and compared the accuracy with individual classifiers such as k-nearest neighbours, support vector machines perceptron, and AdaBoost.

3.1.5 Ectopic Pregnancy

In [16], the authors proposed a decision support system that helps the clinicians to decide initial treatment to be followed for ectopic pregnancy. Algorithms such as multilayer perceptron, deep learning, support vector machine, and naive Bayes were used individually as well as an ensemble of three-stage classifiers to decide the better treatment. The dataset included 406 cases of tubal ectopic pregnancies, with personal and medical variables as well as two-dimensional transvaginal ultrasound. The best accuracy of 96% was obtained when using a support vector machine.

3.1.6 Impact of Dietary Intake

Another important factor we need to consider in pregnant women health is dietary intake. One such study [6] explored the relationship between dietary intake and adverse pregnancy outcomes, where the authors took data of 7572 women from another study [20]. A food frequency questionnaire was used to assess the average daily prenatal consumption of total fruits and vegetables. Using multivariable logistic regression and super learner with targeted maximum likelihood estimation, the authors computed the marginal risk of preterm delivery, preeclampsia, small for gestational age, and gestational diabetes.

3.1.7 Relation Between Particulate Matter$_{10}$(PM_{10}) and Risk of Congenital Heart Disease (CHD)

In [49], the authors employed machine learning models to study the non-linear effects of PM_{10} exposure on the risk of CHDs throughout the key time window of weeks 3–8 in pregnancy. Random forest and gradient boosting were used to examine the association between PM_{10} exposure and the risk of CHDs utilizing demographic data. Dataset includes 39,053 live-born children in Beijing with features maternal nation, gestation age, mother's nation, season of conception, etc.

3.2 Unsupervised Learning

Unsupervised learning such as dimensionality reduction, clustering, and association rule mining were utilized to classify foetal heart rate tracings and foetal health status and predict miscarriage.

3.2.1 Clustering

Clustering is one of the primary unsupervised learning methods used in maternal data. Diff rent types of clustering techniques have been successfully applied to categorize the pregnancy outcomes. For example, a real-time miscarriage prediction using big data was proposed in [4], where k-means clustering was used. A comparative study on different clustering algorithms for real-time miscarriage prediction was proposed by [3]. The data was collected using wearable sensors and mobile applications in every 60 s. Around 100,000 documents with 15 attributes were used for the study. k-means clustering, bisecting k-means, and Gaussian mixture were adopted and reported that k-means and bisecting k-means were more efficient.

A multi-channel approach [39], where electromyograph signals are recorded by using a 4X4 electrode matrix connected to women's abdomen, and an unsupervised learning technique were adopted for the categorization of signals. After extracting relevant features, it was fed into a competitive neural network. Clustering based on correlation was performed to group similar patterns. In a particular study, [51] proposed a data mining software architecture that would aid in the personal monitoring of the pregnancy process in the health sector. Various sorts of data, such as age, gender, location, education, physical attributes, lifestyle habits, and medical history of people, are recorded online by health institutions. Such data can be mined for identifying significant relations and for predictions. The proposed method adopts random forest for classification, k-means for clustering, and association rules on a synthetic dataset, which gave a performance accuracy of 80% on a classification of healthy and risky pregnancy.

A clustering of 3601 uterine contractions based on electrohysterography was performed in a particular work [13], where hierarchical clustering and partitional methods such as K-means and K-medoids clustering were evaluated. The average linkage method was selected for clustering analysis since it presented distinct spectral peaks and outperformed k-means and k-medoids. The study led to spectral characterization of events in uterine. A novel method based on clustering and principal component analysis to extract single-lead noninvasive foetal electrocardiogram was proposed in a study [57].

3.2.2 Dimensionality Reduction

Since many of the datasets used for maternal monitoring include several clinical features, it is important to identify the parameters that are most relevant to the predictive task and visualization. Linear and non-linear methods are used as a preprocessing step in a system that predicts the risk of pregnancy. In [28], principal component analysis was used to identify dimensions in predicting inflammation risk factors for preterm birth. In a preeclampsia prognosis system [21] with 222 samples of normal pregnant woman data and 161 women with preeclampsia, principal component analysis followed by logistic regression wad applied. Principal component analysis and artificial neural networks were used in [38] to study embryo implantation potential.

3.3 Artificial Neural Network

A self-organizing multilayer neural network is proposed in [24] to predict preterm birth using an open dataset with 38 preterm birth and 262 normal birth examples. Artificial neural network and backpropagation algorithm are applied in [35] for predicting four types of pregnancy disorders utilizing 172 patient records. A network design with 17 input layers, 50 hidden layers, and 5 output layers was adopted for the classification task. A proposal of the modular neural network [25] with error back propagation algorithm to classify foetal state using cardiotocography data utilizing 2126 foetal CTG recording, where a 100% classification was achieved for pathogenic cases, while the suspect classification rate was between 90.65 and 98.46% and more than 99% classification rate was achieved for the normal case. Data classification system using an ANN [53] on the same dataset could enhance the performance by classifying pathogenic and normal cases with an unexpected poor result for the suspicious cases.

3.4 Genetic Algorithm

Lu et al. in [34] proposed ensemble machine learning model based on the genetic algorithm to predict foetal weight at different gestational ages. The authors used a dataset with 4212 samples where a cubic spline was used to fit the data extracted from ultrasound reports. Ensemble model consisting of random forest, XG-Boost, and LightGBM algorithms was used to train the model, whereas genetic algorithm was applied for parameter optimization. A novel neuro-fuzzy based on genetic algorithm is proposed in another work [27], to differentiate the foetus growth as abnormal or normal. The method achieved an accuracy of 97%. The optimum feature subset to predict preterm labour was achieved using genetic algorithm along with an ensemble logistic regression, linear discriminant analysis, and k-nearest neighbour classifiers in the study [41]. In this research, an F1 score of 92.04±2.97% was reported on experiment with 326 EHG signals.

3.5 Deep Learning

Recently, deep learning methods are widely applied in medical image analysis because of their improved performance in very large datasets. Since the most commonly used convolutional neural network approaches can automatically extract features without any human intervention, it is preferred over conventional machine learning algorithms. This is effectively applied in foetal segmentation and placental segmentation [12]. In morphologic pattern prediction [37], the authors developed a model to differentiate and classify the existence or non-existence of multiclass morphologic patterns. An accuracy of 88.02% and an F score of 0.8508 in average were achieved by the model on CTG dataset. Another work [55] contributed a deep neural network that precisely detects foetal QRS complexes from foetal ECG. The authors extended the two-dimensional CNN model [59] with three convolutional layers to extract features from single-channel foetal ECG signals by formulating the residual network architecture. An encoder–decoder-based convolutional neural network (CNN), named FetalNet [45] for simultaneous segmentation and classification of the foetal body parts utilizing 700 two-dimensional foetal video sequences from 700 different patients, was proposed in a study. It included eight convolutional blocks, four in the encoder part and four in the decoder part, where encoders were concatenated with attention gates [42] to decoders through skip connections. The study resulted in an accuracy of 0.975.

Further, a pregnancy outcome prediction in the case of patient with a history of recurrent reproductive failure (RRF) was done using deep learning in [22]. The electronic medical record data of 561 patients was the input to an iterative shrinkage and thresholding algorithm with coordinate descent. While Relu and Tanh functions were used in hidden layer, softmax aided the output layer. In this chapter, a model was developed for the prediction of the pregnancy outcomes at any gestational period. The accuracy ranged from 54.2% to 89.7% for each stage in testing data. A deep learning regression model [30] used foetal MRI as the input data and predicted gestational age, where 184 T2-weighted MRI acquisitions from 184 foetuses were considered.

A convolutional neural network (CNN) model based on a weighted voting mechanism that classified the foetal heart rate as normal or pathological state was proposed by [32], which assists obstetrician to make accurate medical decisions. Here imbalanced data was handled using the multi-model training method based on down-sampling algorithm. Analysis of the model using 552 samples from CTU-UHB open database resulted in 0.9 AUC. Foetal hypoxia is a serious condition in which foetus deprives of sufficient oxygen. In [58], the authors proposed a novel computer-aided diagnosis system with an advanced deep learning to predict foetal hypoxia. Using recurrence plot, the 1D time series was transformed into 2D image

Table 1 Summary of AI techniques applied in maternal care, prediction tasks, and performance

Prediction task	AI techniques	Metrics and score	Reference
Preterm birth	SVC, k-nn, decision tree	95% AUC	[15]
Miscarriages	k-NN	98% accuracy	[29]
Foetal health	Random forest	99.2% accuracy	[50]
Preeclampsia	Stochastic gradient descent	97.3% accuracy	[26]
Ectopic pregnancy	Multilayer perceptron, SVM	96.1% accuracy	[16]
Foetal health	Decision forest	89.5% accuracy	[1]
Foetal health	RF,K-Means	80% accuracy	[51]
QRS complex	K-Means	95.78% accuracy	[57]
Foetal health	ANN	98.46% accuracy	[25]
Foetal growth	Ensembling	64.5% accuracy	[34]
Foetal growth	Neuro-fuzzy, genetic algorithm	97% accuracy	[27]
Foetal health	SVM, ANN with wrapper class	F1 92.04	[11]
Morphological pattern	Deep neural network	88.02% accuracy	[37]
Pregnancy outcome	Deep neural network	89.7% accuracy	[22]

and designed CNN model to learn the intrinsic patterns automatically. An accuracy of 98.69% was achieved on a dataset of 552 intrapartum CTG recordings from a subset of 9164 recordings. Table 1 provides a summary of AI techniques applied in various papers reviewed during this study. While Table 2 provides a summary of various data sources utilized in the research works.

4 Conclusion and Future Research

In this chapter, we presented a summary of different machine learning techniques applied in pregnancy risk prediction. Continuous monitoring and pregnancy outcome prediction frameworks provide high-quality healthcare for women in the prevention of complications related to pregnancy. These models utilize either historical or real-time data from the patients and generate the prediction models. The system acts as an individual self-management system that seamlessly performs the prognosis and recommends the required type of exercise, nutrition intake, and clinical support.

Storing, processing, and transferring large volumes of remote health data over the network becomes a critical challenge in these types of decision support systems. In addition to the huge power consumption, data analysis at the server demands the bandwidth of resource-limited networks. Although cloud-based systems solve data storage problems, response time is comparatively long and real-time patient monitoring becomes impractical. In such cases, edge computing can be utilized to perform data processing on the peripheral devices themselves. The design of lightweight machine learning algorithms facilitates on-device learning and faster decision-making. Combining learning based on historical data and real-time data facilitates better prediction results.

Table 2 Summary of data sources

SI No	Data source	Description	Papers used
1	TPEHG dataset in Physionet	300 EHG data	[15], [55], [41]
2	UCI Machine Learning repository	2126 Cardiotocography data	[37], [50], [37], [11]
3	Early Pregnancy Units, Imperial College Healthcare Trust, London, UK	79 USG images	[29]
4	Physionet/computing challenge database	12 lead ECG	[55]
5	EMR data Fertility Center of Shenzhen Zhongshan Urology Hospital, China	992 samples	[22]
6	Physionet CTU-CHB Intrapartum Cardiotocography Database	552 samples	[17], [58]
7	Data on tubal ectopic pregnancies	406 samples	[16]
8	CTG recordings from University Hospital in Brno Czech Republic	552 samples	[14]
9	PhysioNet/CinC Challenge 2013 Set A	75 samples	[57]
10	Yonsei University Healthcare Center in Seoul	11,006 samples	[26]
11	EMR data from Fertility Center of Shenzhen Zhongshan Urology Hospital, China	561 samples	[22]
12	Physionet cardiology challenge database	75 samples	[55]
13	Foetal dataset	700 2D foetal video	[45]
14	Open database CTU-UHB	552 CTG recordings	[32]

References

1. A. Akbulut, E. Ertugrul, V. Topcu, Fetal health status prediction based on maternal clinical history using machine learning techniques. Comput. Methods Prog. Biomed. **163**, 87–100 (2018)
2. A. Arunan, R.K. Pathinarupothi, M.V. Ramesh, A real-time detection and warning of cardiovascular disease LAHB for a wearable wireless ECG device, in *2016 IEEE-EMBS International Conference on Biomedical and Health Informatics (BHI)* (IEEE, Piscataway, 2016), pp. 98–101
3. H. Asri, Big data and IoT for real-time miscarriage prediction: a clustering comparative study. Proc. Comput. Sci. **191**, 200–206 (2021)
4. H. Asri, H. Mousannif, H. Al Moatassime, Real-time miscarriage prediction with spark. Proc. Comput. Sci. **113**, 423–428 (2017)
5. C.B. Benson, P.M. Doubilet, The history of imaging in obstetrics. Radiology **273**(2S), S92–S110 (2014)

6. L.M. Bodnar, A.R. Cartus, S.I. Kirkpatrick, K.P. Himes, E.H. Kennedy, H.N. Simhan, W.A. Grobman, J.Y. Duffy, R.M. Silver, S. Parry, et al., Machine learning as a strategy to account for dietary synergy: an illustration based on dietary intake and adverse pregnancy outcomes. Am. J. Clin. Nutr. **111**(6), 1235–1243 (2020)
7. N.V. Chawla, K.W. Bowyer, L.O. Hall, W.P. Kegelmeyer, SMOTE: synthetic minority over-sampling technique. J. Artif. Intell. Res. **16**, 321–357 (2002)
8. CTG, https://archive.ics.uci.edu/ml/datasets/cardiotocography. Accessed 12 April 2021
9. CTG-UHB, https://physionet.org/content/ctu-uhb-ctgdb/1.0.0/. Accessed 12 April 2021
10. Depression, https://americanpregnancy.org/healthy-pregnancy/pregnancy-health-wellness/depression-during-pregnancy/. Accessed 27 Nov 2021
11. T.D. Deressa, K. Kadam, Knowledge management system for fetal movement during pregnancy. Int. J. Pure Appl. Math. **118**(24) (2018)
12. P.H. Diniz, Y. Yin, S. Collins, Deep learning strategies for ultrasound in pregnancy. Eur. Med. J. Reprod. Health **6**(1), 73 (2020)
13. F. Esgalhado, A.G. Batista, H. Mouriño, S. Russo, C.R.P. Dos Reis, F. Serrano, V. Vassilenko, M. Ortigueira, Uterine contractions clustering based on electrohysterography. Comput. Biol. Med. **123**, 103897 (2020)
14. G. Feng, J.G. Quirk, P.M. Djurić, Supervised and unsupervised learning of fetal heart rate tracings with deep Gaussian processes, in *2018 14th Symposium on Neural Networks and Applications (NEUREL)* (IEEE, Piscataway, 2018), pp. 1–6
15. P. Fergus, P. Cheung, A. Hussain, D. Al-Jumeily, C. Dobbins, S. Iram, Prediction of preterm deliveries from EHG signals using machine learning. PloS One **8**(10), e77154 (2013)
16. A.D.R. Fernández, D.R. Fernández, M.T.P. Sánchez, A decision support system for predicting the treatment of ectopic pregnancies. Int. J. Med. Infor. **129**, 198–204 (2019)
17. G. Georgoulas, J. Spilka, P. Karvelis, V. Chudáček, C. Stylios, L. Lhotská, A three class treatment of the FHR classification problem using latent class analysis labeling, in *2014 36th Annual International Conference of the IEEE Engineering in Medicine and Biology Society* (IEEE, Piscataway, 2014), pp. 46–49
18. Gestational diabetes, http://www.acog.org/Patients/FAQs/Gestational-Diabetes. Accessed 27 Nov 2021
19. R.L. Goldenberg, J.F. Culhane, J.D. Iams, R. Romero, Epidemiology and causes of preterm birth. Lancet **371**(9606), 75–84 (2008)
20. D.M. Haas, C.B. Parker, D.A. Wing, S. Parry, W.A. Grobman, B.M. Mercer, H.N. Simhan, M.K. Hoffman, R.M. Silver, P. Wadhwa, et al., A description of the methods of the nulliparous pregnancy outcomes study: monitoring mothers-to-be (nuMoM2b). Am. J. Obstet. Gynecol. **212**(4), 539.e1–539.e24 (2015)
21. Y. Hou, X. Rui, L. Yun, J. Lin, A15411 a risk prediction model for preeclampsia in the Chinese Han pregnant women. J. Hypertens. **36**, e230 (2018)
22. C. Huang, Z. Xiang, Y. Zhang, D.S. Tan, C.K. Yip, Z. Liu, Y. Li, S. Yu, L. Diao, L.Y. Wong, et al., Using deep learning in a monocentric study to characterize maternal immune environment for predicting pregnancy outcomes in the recurrent reproductive failure patients. Front. Immunol. **12** (2021)
23. K.J. Hurt, M.W. Guile, J.L. Bienstock, H.E. Fox, E.E. Wallach, et al., *The Johns Hopkins Manual of Gynecology and Obstetrics* (Lippincott Williams & Wilkins, Philadelphia, 2012)
24. A.J. Hussain, P. Fergus, H. Al-Askar, D. Al-Jumeily, F. Jager, Dynamic neural network architecture inspired by the immune algorithm to predict preterm deliveries in pregnant women. Neurocomputing **151**, 963–974 (2015)
25. S. Jadhav, S. Nalbalwar, A. Ghatol, Modular neural network model based foetal state classification, in *2011 IEEE International Conference on Bioinformatics and Biomedicine Workshops (BIBMW)* (IEEE, Piscataway, 2011), pp. 915–917
26. J.H. Jhee, S. Lee, Y. Park, S.E. Lee, Y.A. Kim, S.W. Kang, J.Y. Kwon, J.T. Park, Prediction model development of late-onset preeclampsia using machine learning-based methods. PLoS One **14**(8), e0221202 (2019)

27. P. Kaur, G. Singh, P. Kaur, An intelligent validation system for diagnostic and prognosis of ultrasound fetal growth analysis using neuro-fuzzy based on genetic algorithm. Egypt. Inform. J. **20**(1), 55–87 (2019)
28. L.S. Keenan-Devlin, M. Caplan, A. Freedman, K. Kuchta, W. Grobman, C. Buss, E.K. Adam, S. Entringer, G.E. Miller, A.E. Borders, Using principal component analysis to examine associations of early pregnancy inflammatory biomarker profiles and adverse birth outcomes. Am. J. Reprod. Immunol. (2021)
29. S. Khazendar, J. Farren, Al-Assam, H., H. Du, A. Sayasneh, T. Bourne, S.A. Jassim, Automatic identification of miscarriage cases supported by decision strength using ultrasound images of the gestational sac. Ann. BMVA **2015**(5), 1–16 (2015)
30. Y. Kojita, H. Matsuo, T. Kanda, M. Nishio, K. Sofue, M. Nogami, A.K. Kono, M. Hori, T. Murakami, Deep learning model for predicting gestational age after the first trimester using fetal MRI. Eur. Radiol. **31**(6), 3775–3782 (2021)
31. L. Leeman, P. Fontaine, Hypertensive disorders of pregnancy. American Family Physicina (serial online) 2008 (visitado 2011 dic 3)
32. S. Liang, Q. Li, Automatic evaluation of fetal heart rate based on deep learning, in *2021 2nd Information Communication Technologies Conference (ICTC)* (IEEE, Piscataway, 2021), pp. 235–240
33. L. Liu, Y. Jiao, X. Li, Y. Ouyang, D. Shi, Machine learning algorithms to predict early pregnancy loss after in vitro fertilization-embryo transfer with fetal heart rate as a strong predictor. Comput. Methods Prog. Biomed. **196**, 105624 (2020)
34. Y. Lu, X. Zhang, X. Fu, F. Chen, K.K. Wong, Ensemble machine learning for estimating fetal weight at varying gestational age, in *Proceedings of the AAAI Conference on Artificial Intelligence*, vol. 33 (2019), pp. 9522–9527
35. D.S. Maylawati, M.A. Ramdhani, W.B. Zulfikar, I. Taufik, W. Darmalaksana, Expert system for predicting the early pregnancy with disorders using artificial neural network, in *2017 5th International Conference on Cyber and IT Service Management (CITSM)* (IEEE, Piscataway, 2017), pp. 1–6
36. B.E. Metzger, D.R. Coustan, O. Committee, et al., Summary and recommendations of the fourth international workshop-conference on gestational diabetes mellitus. Diabetes Care **21**, B161 (1998)
37. J.H. Miao, K.H. Miao, Cardiotocographic diagnosis of fetal health based on multiclass morphologic pattern predictions using deep learning classification. Int. J. Adv. Comput. Sci. Appl. **9**(5), 1–11 (2018)
38. R. Milewski, A. Kuczyńska, B. Stankiewicz, W. Kuczyński, How much information about embryo implantation potential is included in morphokinetic data? A prediction model based on artificial neural networks and principal component analysis. Adv. Med. Sci. **62**(1), 202–206 (2017)
39. B. Moslem, M.O. Diab, M. Khalil, C. Marque, Classification of multichannel uterine EMG signals by using unsupervised competitive learning, in *2011 IEEE Workshop on Signal Processing Systems (SiPS)* (IEEE, Piscataway, 2011), pp. 267–272
40. NFHS, https://ruralindiaonline.org/en/library/resource/national-family-health-survey-nfhs-4-2015-16-india/. Accessed 27 Nov 2021
41. F. Nieto-del Amor, G. Prats-Boluda, J.L. Martinez-De-Juan, A. Diaz-Martinez, R. Monfort-Ortiz, V.J. Diago-Almela, Y. Ye-Lin, Optimized feature subset selection using genetic algorithm for preterm labor prediction based on electrohysterography. Sensors **21**(10), 3350 (2021)
42. O. Oktay, J. Schlemper, L.L. Folgoc, M. Lee, M. Heinrich, K. Misawa, K. Mori, S. McDonagh, N.Y. Hammerla, B. Kainz, et al., Attention U-Net: learning where to look for the pancreas. Preprint. arXiv:1804.03999 (2018)
43. R.K. Pathinarupothi, M.V. Ramesh, E. Rangan, Multi-layer architectures for remote health monitoring, in 2016 IEEE 18th International Conference on e-Health Networking, Applications and Services (Healthcom) (IEEE, Piscataway, 2016), pp. 1–6

44. R.K. Pathinarupothi, E. Rangan, B. Alangot, M.V. Ramesh, RASPRO: rapid summarization for effective prognosis in wireless remote health monitoring, in *2016 IEEE Wireless Health (WH)* (IEEE, Piscataway, 2016), pp. 1–6
45. S. Płotka, T. Włodarczyk, A. Klasa, M. Lipa, A. Sitek, T. Trzciński, FetalNet: multi-task deep learning framework for fetal ultrasound biometric measurements. Preprint. arXiv:2107.06943 (2021)
46. Preeclampsia, http://www.acog.org/Patients/FAQs/Preeclampsia-and-High-Blood-Pressure-During-Pregnancy. Accessed 27 Nov 2021
47. A. Priya, S. Chaturvedi, S.K. Bhasin, M.S. Bhatia, G. Radhakrishnan, Depression, anxiety and stress among pregnant women: a community-based study. Indian J. Psychiatry **60**(1), 151 (2018)
48. M.V. Ramesh, S. Anand, P. Rekha, A mobile software for health professionals to monitor remote patients, in *2012 Ninth International Conference on Wireless and Optical Communications Networks (WOCN)* (IEEE, Piscataway, 2012), pp. 1–4
49. Z. Ren, J. Zhu, Y. Gao, Q. Yin, M. Hu, L. Dai, C. Deng, L. Yi, K. Deng, Y. Wang, et al., Maternal exposure to ambient PM10 during pregnancy increases the risk of congenital heart defects: evidence from machine learning models. Sci. Total Environ. **630**, 1–10 (2018)
50. H. Sahin, A. Subasi, Classification of the cardiotocogram data for anticipation of fetal risks using machine learning techniques. Appl. Soft Comput. **33**, 231–238 (2015)
51. Y. Santur, S.G. Santur, Knowledge mining approach for healthy monitoring from pregnancy data with big volumes. Int. J. Intell. Syst. Appl. Eng., 141–145 (2016)
52. O. Simbolon, M.N. Widyawati, K. Kurnianingsih, N. Kubota, N. Ng, Predicting the risk of preeclampsia using soft voting-based ensemble and its recommendation, in *2020 International Symposium on Community-centric Systems (CcS)* (IEEE, Piscataway, 2020), pp. 1–6
53. C. Sundar, M. Chitradevi, G. Geetharamani, Classification of cardiotocogram data using neural network based machine learning technique. Int. J. Comput. Appl. **47**(14) (2012)
54. TPEHG, https://physionet.org/content/tpehgdb/1.0.1/. Accessed 12 April 2021
55. K. Vo, T. Le, A.M. Rahmani, N. Dutt, H. Cao, An efficient and robust deep learning method with 1-d octave convolution to extract fetal electrocardiogram. Sensors **20**(13), 3757 (2020)
56. S. Wanriko, N. Hnoohom, K. Wongpatikaseree, A. Jitpattanakul, O. Musigavong, Risk assessment of pregnancy-induced hypertension using a machine learning approach, in *2021 Joint International Conference on Digital Arts, Media and Technology with ECTI Northern Section Conference on Electrical, Electronics, Computer and Telecommunication Engineering* (IEEE, Piscataway, 2021), pp. 233–237
57. Y. Zhang, S. Yu, Single-lead noninvasive fetal ECG extraction by means of combining clustering and principal components analysis. Med. Biol. Eng. Comput. **58**(2), 419–432 (2020)
58. Z. Zhao, Y. Zhang, Z. Comert, Y. Deng, Computer-aided diagnosis system of fetal hypoxia incorporating recurrence plot with convolutional neural network. Front. Physiol. **10**, 255 (2019)
59. W. Zhong, L. Liao, X. Guo, G. Wang, A deep learning approach for fetal QRS complex detection. Physiol. Meas. **39**(4), 045004 (2018)

Detection of Non-Technical Losses in Power Utilities Using Machine Learning

Shaima Al-Maskari and K. Vijayalakshmi

Abstract Non-technical losses are one of the issues electricity companies face and lead to a significant source of revenue losses. The aim of this study is to use the machine learning algorithms to reach meaningful patterns and enable the businesses to make proactive, knowledge-driven decisions. This research work is proposed to detect the non-technical losses by getting the list of customers according to their probability of having an anomaly in their electricity meter using machine learning algorithms. In this study, many research papers on non-technical losses are reviewed, and their key features are summarized. Data collection is done from one of the leading electricity distribution companies in Oman, and it had monthly electricity consumption from May 2020 to May 2021 of more than 400,000 accounts. This study uses 284,737 normal and anomaly accounts of residential and commercial customers to train and test the ML algorithms. Thirty-four features are extracted from the raw data, and ten top features are selected to train and test the machine learning models. The data is modeled using three supervised learning algorithms, namely, XgBoost, ANN, and KNN. The results showed that ANN outperformed the other two algorithms when using the SMOTE method with train/test split with 85% of AUC score and 73% of recall score. Finally, the ANN algorithm using SMOTE is used to get a list of customers with an estimated probability of having anomalies in the meter.

Keywords Non-technical losses (NTL) · Electricity consumption · Anomaly · ANN · XgBoost · KNN

S. Al-Maskari (✉) · K. Vijayalakshmi
National University of Science and Technology, Muscat, Sultanate of Oman
e-mail: sh.almaskeri@gmail.com; vijayalakshmi@nu.edu.om

A. Haldorai et al. (eds.), *4th EAI International Conference on Big Data Innovation for Sustainable Cognitive Computing*, EAI/Springer Innovations in Communication and Computing, https://doi.org/10.1007/978-3-031-07654-1_4

1 Introduction

In recent years, the global electricity demand has been growing. This growth raises economic and regulatory concerns that are directly related to the sustainability of distribution systems as not all the electricity made on hand reaches the end customers despite the growing demand. A significant amount of energy is lost through the distribution system, and there are two types of losses, technical and non-technical losses (NTL). Technical losses are dissipation of electricity that naturally happens in transportation, transformation and distribution, and energy measurement. Non-technical losses are non-natural losses that are consumed but not invoiced, and this occurs due to either error in the metering or billing or the illegal connection due to the non-legitimate behavior of consumers [1]. Technical losses are determined easily by using only two parameters: the total load on the grid and the total bill of the consumed energy. On the other hand, NTL is difficult to predict, and the main priority of any power distribution company (PDC) is minimizing the impact due to NTL [2]. There are various impacts of the NTL, and the most important is the loss of revenue. It not only results in considerable revenue losses, but it also has an impact on the operation of the power system since it provides uncertainty of the actual consumption [3]. In terms of economy, the NTL takes 10–40% of the entire capacity of the electricity company's economy [4]. This affects the electricity company's profit by overloading the generators, and additional electricity capacity is required to face unexpected changes in distribution. At the same time, it affects the performance of the appliances which causes variations in voltage because of the overload. NTL causes financial issues for the companies who failed to manage the losses and decreases legally collected tariff which implies that non-fraudulent customers are paying for fraudulent customers. In society terms, illegal connections and manipulation of the network or meter tampering are harmful not only for who cause this but also for the technicians who maintain the installation and people who are living in the building [4] [5]. In government terms, NTL negatively affects the investment decision to expand connectivity and electricity access to undeveloped regions [6].

Many research projects of non-technical losses in energy sector are in their initial stage and are not fully effective in Oman. In order to detect NTLs, electricity companies used to conduct a metering device inspection based on observations gathered from a random evaluation of consumer billing profiles. The success rate of such random inspections is quite low because the behavior of the consumers' consumption pattern is not considered during the detection process. This method is ineffective and inefficient because it does not check all bills, and it is very costly and time-consuming [2]. Smart meters (SMs) give electricity companies the ability to come up with new and creative techniques to identify NTL through analyzing data collected by smart meter measurement systems. Electricity companies now have access to frequent EC measurements because of using SM, providing them a greater insight of their customers' consumption behavior [3]. The value of this technology is due to the increase of efficiency of estimation algorithms and classification of users [7].

The goal of this research is to align electricity companies with Oman's 2040 vision. The aim is to achieve energy security through renewable energy, diversified sources of energy, and rationalized consumption. The annual report of the electricity distribution company indicated that the company has achieved a sustainable reduction in distribution losses from 17.58% to 7.76% from 2009 to 2020. The aim of the company is to reach 5% losses by using the new technologies that enhance the way of detecting the losses. This research aims to study the effectiveness of using machine learning algorithms to detect non-technical losses. The result of this study can be used by the electricity companies to reduce the wastage of resources. This study can help the electricity companies to fight against the non-technical losses by studying usage data from a large group of customers and highlighting anomalies in meter readings. The scope of the study is to cover Muscat governorate and can be expanded to other governorates in Oman.

2 Literature Review

Non-technical losses detection is one of the major concerns of the distribution system operators. Numerous studies have been conducted in the literature to reduce the effect on the economy and sustainability of the power utilities. Two types of solutions exist in the literature for NTL detection which are hardware and non-hardware-based solutions. As a result, this research proposed a non-hardware-based solution for NTL detection. Therefore, the most recent development of the NTL detection techniques has been reviewed and presented in this research work to emphasize the significance of the current study.

2.1 Feature Selection

It is common to use not only the raw data but also features extracted from that data to reduce the classifier performance. These features are calculated from the time series data of the customer and specifically from the active consumption curve. Reducing the size of the dataset may improve the classification metrics and provide a level of anonymization. A lot of features can be extracted from the time series and that are related to the NTL detection should be considered [6].

Many feature selection algorithms are used to defined optimal set of features. In [4] the authors used LinearSVC, SW-F, and RF-RFE as feature selection techniques to determine the optimal features to be considered for the classification stage. Pearson's chi-square (X2) feature selection method was used in [2] to select the relevant features from the dataset. Authors in [3] used several type of features which need different types of feature selection algorithms. They used quality byte (QB) measures to extract features related to alarms. Also, they used Z-score to detect anomalous measurements, and they used local outlier factor (LOF) to detect an unusual customer behavior. Authors in Ghori et al. in [8] used F-measure and Gini Index to find the top k number of F-measure features. A combination of genetic programing and finite mixture model clustering model was used by the authors [9] to create a set of features that describe the dynamic of the demand over time to easily detect anomalies and fraud.

2.2 *Imbalance Dataset*

The imbalance issue happens in classification algorithms when one class is disproportionately larger than another. For the imbalance dataset in the training process, some of the algorithms favor the positive data and ignore the important data in a few negative data. Optimizing the unbalance dataset plays an important role in increasing the algorithm's efficiency and accuracy [10]. To solve this issue many studies have been done in various ways [5]. isolated the overrepresentation of types of customers and generated specific campaigns for them to avoid the system to learn to classify that type of customer. Also, for the underrepresentation where the information of the customers are not good enough to detect the fraud, they were isolated to force the system to detect the NTL pattern for this type of customers [3]. used under-sampling techniques to reduce the honest customers during the training from the imbalance data. They did it by using two ways, first one by removing the customers who did not classify as anomalies in the smart meter and have been misclassified by the inspectors more than three times. The second way is by removing the honest customers using a random seed with a different number [10]. proposed a new algorithm called K-SMOTE which combines the K-means algorithm with SMOTE (Synthetic Minority Oversampling Technique) algorithm to oversample the electricity data which has only 15.84% rate of data repetition.

2.3 *Machine Learning Algorithms to Detect NTL*

Many literatures exist for implementation of the machine learning solutions for detecting non-technical losses in the energy sector with different approaches. It depends on the type of algorithm used (supervised, unsupervised) or on the type of data that is used in order to reduce the effect on the economy and sustainability of the power utilities.

Puig and Carmona in [5] created binary classification supervised models by using gradient boosting on decision tree (GBM) libraries such as XGBoost, LightGBM, and CatBoost. The study means the paper in [5] extracted the features of the NTL from historical campaigns (on-field inspections) for 5 years. They achieved up to 36% of precision in campaigns to detect NTL detection for customers with contracts and 50% for customers with no contract whose contract is resigned but the smart meter installation is not removed. The limitation of this study is the manual analysis of the system and decided if the correlation of important features is because of causal pattern or derived from biased labeled instances.

Dominguez et al. in [4] tried to recognize NTL using several machine learning algorithms. They used on-field inspections data of Home and commercial accounts to validate if the account has non-technical losses or not. They used nine classifiers which are KNN, SVM, LDA and QLDA, DT, RF, NB, GNB, GBM, and ADB, to train several subsets that are provided by the three feature selection methods to find the one that maximizes the accuracy. They found that the tree-based method (GBM) and adaptive boosting (ADB) outstand other methods.

Saeed et al. in [2] used Boosted C5.0 decision tree (DT) algorithm to classify the honest and fraudster customers, and it is trained by using real data of domestic and commercial accounts. Boosted C5.0 DT classifier was compared with the state-of-the-art machine learning algorithms (SVM, random forest, ANN, and XGBoost tree). The final result shows that the Boosted C5.0 decision tree outperforms the other algorithms, and it classifies 98% of the honest and 78% of the fraudster customers accurately.

Qu et al. in [10] used random forest algorithm to detect the electricity theft behavior, and they used K-SMOTE which is based on K-mean clustering and SMOTE method to avoid the influence of the imbalanced data. They used short-term load data at intervals of 30 min through a smart meter of 60 urban electricity users. They found that the proposed method is better than the conventional detection methods (BPN and SVM) in higher accuracy and stronger stability.

Ghori et al. in [8] used real data of consumer's monthly consumption records over a period of 15 months to evaluate 15 machine learning classifiers across 9 types for non-technical loss detection. They found that the ensemble methods outperform other types with respect to F-measure and ANN outperform other types with respect to Recall. For the individual classifier analysis, CatBoost outperforms other classifiers with respect to F-measure with approximately 98.5%, and at the same time, the MLP classifier outperforms others with respect to the recall with approximately 99.4%. The authors recommend to create a benchmark datasets that are widely used in NTL detection. They also recommended to use the penalized machine learning models with weighted classifiers.

[9] proposed a feature selection algorithm that combined the mixture model, clustering model, and an evolutionary genetic programming algorithm to create a set of features that describe the dynamic of the demand over time to easily detect anomalies and fraud. They used half-hourly energy consumption data of more than 4000 households for 18 months to train the classification algorithms. They evaluate different machine learning algorithms, GBM, SVM, RF, NN, and the KNN algorithm. When dealing with actual attack conditions such as unseen attacks and irregular attacks and zero-day attacks and dealing with theft events, the proposed method was also proved to increase performance.

Punmiya and Choe in [11] did a comprehensive comparison between the most recent GBCs including extreme gradient boosting (XGBoost), categorical boosting (CatBoost), and light gradient boosting method (LightGBM), and based on these three algorithms, a gradient boosting theft detector (GBTD) was proposed. GBTD had a feature engineering-based preprocessing module to improve the detection rate (DR), false-positive rate (FPR), and time complexity. The authors used a half-hourly electricity consumption of 420 days, and they used 48 features that describe customers' daily usage. The authors generated theft accounts using six cases that generate a more practical real-time malicious behavior pattern. The final result shows that LightGBM and CatBoost outperform the XGBoost in terms of detection rate. LightGBM has the highest FPR, and CatBoost is the lowest. The study proves that the proposed algorithm minimizes FPR and reduces processing time and data storage space.

3 Methodology

3.1 Data Flow

The generalized methodology covers importing data, cleansing, preprocessing, feature extraction and selection, mining, prediction accuracy, and visualization as shown in Fig. 1.

3.2 Data Collection

The electricity consumption data of this study is provided by one of the leading Electricity distribution companies in Oman which is responsible for the distribution and supply of electricity in Sultanate of Oman. The obtained data has three datasets. The first dataset has monthly kWh consumption data of 13 months from May 2020 to May 2021 for each meter and account and the monthly read type (actual or estimate). The second dataset has the customer profile which has tariff type, meter location (latitude and longitude), and zone. The last dataset has the list of anomaly accounts that have the anomalies caused by tampering, meter faulty, and meter stop.

3.3 Data Cleaning and Preprocessing

The data in the real world is frequently incomplete, noisy, and inconsistent. The preprocessing step helps to clean and reduce the data and also to improve the accuracy and reduce the processing time [12].

Three cleaning and preprocessing steps that have been taken are:

- Noisy Removal.

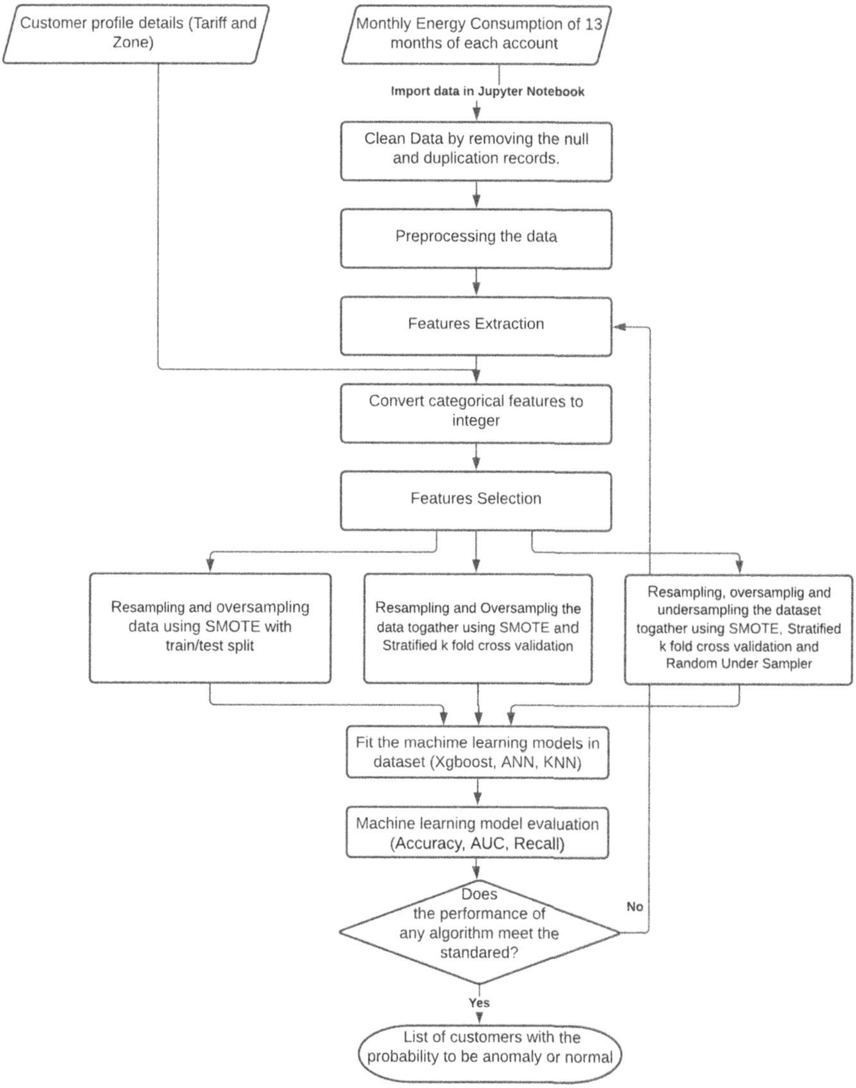

Fig. 1 Data flow diagram

The raw data is expected to have null or zero values. So, there is a need to clean the data before implementing any procedures because there is a chance that it might affect the final result. The used dataset in this study has many duplicated rows for the same smart meter account and for the same month and year and the duplicates are removed.

- Filtration.

The dataset has monthly electricity consumption of 13 months from May 2020 to May 2021. The dataset has more than 400,000 accounts. Some of these accounts do not have complete information for all 13 months. So, the data is filtered to have only the accounts that have full consumption details of 13 months. In addition, the accounts that have more than 11 estimated months are removed from the dataset because this study is dealing with the actual electricity consumption to detect NTL, and it is difficult to calculate the actual electricity consumption if all months are estimated.

3.4 Feature Extraction

The dataset has a huge amount of raw data that is difficult to manage and process. Feature extraction is a process that is used to reduce the raw data in the datasets to more manageable data that can be easily processed. The raw data in the dataset has a lot of variables and features that need a lot of resources to manage and process. Feature extraction can select or combine different variables to make new features that cause reducing the amount of data that need to be processed [13]. After cleaning and preprocessing the datasets, the three fields of the electricity consumption, customer profile, and anomaly accounts are combined together to have one dataset to make the feature extraction process more powerful. There are different types of features that are extracted from the dataset shown in Table 1.

3.5 Monitoring and Prediction

To make predication of the current observation, extracted features are used. The features with high correlation with the anomaly accounts are monitored and analyzed. Then they are used to detect the anomalies in accounts that have similar features using machine learning. The proposed model detect the non-technical losses by estimating the anomaly probability of each accounts.

4 Implementation

4.1 Convert Categorical Features to Binary

Most of the machine learning algorithms require numerical input and output for fitting and evaluating. There are two categorical features that are used in this study, tariff and zone. Using one-hot encoding creates binary columns for each category.

Table 1 Feature extraction

Type of features	Features
Raw consumption: Features that are related to the consumption of kWh during a period of time.	Minimum and maximum consumption during the 13 months; mean; median; standard deviation; total consumption during 13 months; first, second, and third quartile consumption; seasonal consumption (summer, winter, spring and autumn); number of consumption months more than 50,000 kWh, and number of consumption months less than 100 kWh and this range is considered based on the maximum and minimum consumption
Consumption changes: Features that compare through the ratio between customer consumption of two distinct periods of time to discover the changes in the customer's consumption patterns	Consumption ratio between seasons (Summer and Autumn, Autumn and Winter, Winter and Spring, Spring and Summer, and Summer and Winter. Also, because the dataset has the amount of consumption of May 2020 and May 2021, the ratio between two months is included
Consumption curve: The features that compare the customer's consumption and average consumption of the same period	Ratio of customer's consumption with the average consumption of customers in the same zone and the tariff of customer of each season. Ratio of average and the maximum consumption with the customer consumption
Consumption anomalies: Features that calculate the number of outliers in the customer consumption	Number of z-score and IQR outliers
Static features: Features that are static and do not change. This study includes the tariff of the account and zone	Tariff and zone

4.2 Select Features

Thirty-four features are extracted from the energy consumption and customer profile datasets. Instead of using the whole 34 features, there is a need to choose the right subset of features to train the machine learning algorithm faster and to reduce the complexity and improve the accuracy of the machine learning algorithm. Using Pearson's correlation coefficient, the top 10 features that have a high correlation with the anomaly variable are found. Figure 2 shows the top 10 features and shows the number of estimated reading that has the strongest relationship with the anomaly variable.

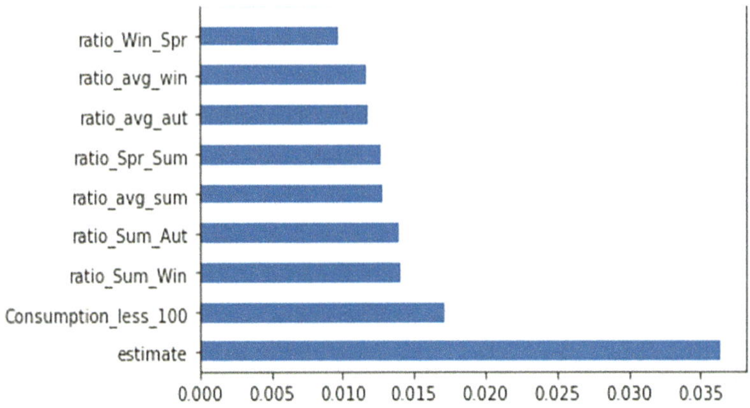

Fig. 2 Top 10 features that have a high correlation with the anomaly variable

Table 2 XgBoost grid search

Hyperparameter	Range	Best
Learning rate	0.01, 0.10	0.01
Maximum depth	7,15	7
Minimum child weight	1,10	10
Gamma	0.2,0.5	0.2
Column sample by tree	0.3, 0.5, 0.7	0.3

4.3 Hyperparameters

Hyperparameters are variables that are not learned directly by estimators. They are supplied as arguments to the constructors of the estimator classes. Grid Search cross-validation considers every possible parameter combination and can locate a suitable parameter combination much faster. A small subset of the parameters can have a significant impact on the model's predictive or computation performance, while the rest can be kept at their default values. Grid Search is applied to the machine learning algorithms that are used in this study to get optimal hyperparameters for each algorithm.

4.3.1 Extreme Gradient Boosted Trees (XGBoost)

XGBoost is a DT-based ensemble algorithm that uses a gradient boosting framework. It creates an additive model by combining the predictions of multiple weak learners [2]. Table 2 shows the hyper parameters used during grid search and the best result.

Table 3 ANN grid search

Hyperparameter	Range	Best
Solver	Lbfgs, sgd, Adam	Lbfgs
Hidden layer sizes	(100,),(50,100,), (50,75,100,)	(100,)
Learning rate	Constant, adaptive, invscaling	Constant

Table 4 KNN grid search

Hyperparameter	Range	Best
Number of neighbors	5,7,9,11,13,15	7
Weights	Uniform, distance	Uniform
Distance metric	Minkowski, euclidean, Manhattan	Minkowski

4.3.2 Artificial Neural Network (ANN)

ANN is a model that is typically used when the information available is insufficient to complete a classification task. ANN has three layers that are all connected. The main layer is made up of input neurons that sends data to the second layer and then to the third layer successively [2]. Table 3 shows the hyperparameters used during grid search and the best result.

4.3.3 K-Nearest Neighbors (KNN)

KNN is one of the simplest supervised classification algorithms for detecting NTL. It finds the closest neighbors using the training data at test time [3]. Table 4 shows the hyperparameters used during grid search and the best result.

4.4 Modeling

The used dataset has 2,84,737 accounts with only 115 accounts classified as anomaly. The dataset is an imbalanced dataset because the number of normal accounts is extremely more than the anomaly accounts. This imbalance issue affects the final result of fitting and evaluating the machine learning algorithm and most of the machine learning algorithms ignore the minority classes. So, to increase the successful anomaly detection, resampling the dataset is required. SMOTE and Random Under Sampler are two commonly used algorithms to solve the imbalanced dataset.

SMOTE is one of the techniques that is used to oversample minority observations. The SMOTE creates synthetic anomaly cases based on the nearest anomaly neighbors' characteristics. Each classification methods need training and testing datasets. The training dataset is used to train the machine learning algorithm, and

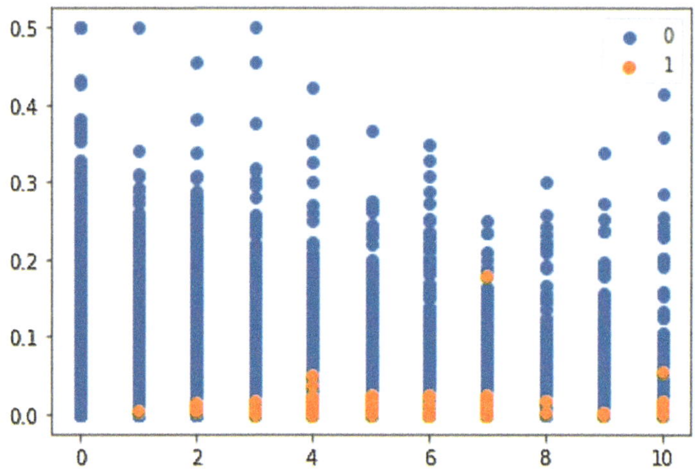

Fig. 3 Before oversampling

the testing is used to test the result of training datasets. This study used two ways of dividing the data, random splitting using the train/test split method by specifying the size of the training and testing datasets. Another way of splitting the data is by using the Stratified K-Folds cross-validator method.

4.4.1 SMOTE Using Train/Test Split

The dataset is converted into two variables, data columns (X) which are independent variables that have all the features that are used to detect the anomaly and label column (Y) which is a dependent variable that shows if the account is an anomaly or not. This study uses 70% of the data for training and the rest 30% of the data for the testing. The training and test sets are initialized for the classification machine learning algorithms used in this study. SMOTE is applied on the training sets using SMOTE function. Figures 3 and 4 show the datasets before and after oversampling in plot scatter charts. Machine learning algorithms with the hyperparameters that are tuned to each of them are trained using the new oversampled datasets, and then they are tested and evaluated using test datasets.

4.4.2 SMOTE with Stratified K-Fold Cross-Validator

Instead of dividing the dataset into training and testing randomly like the previous section method, K-fold cross-validation is used to divide the datasets into k disjoint subsets of approximately equal size. The ML algorithm is trained using K-1 subsets which represent the training set and the ML algorithm is tested using the remaining

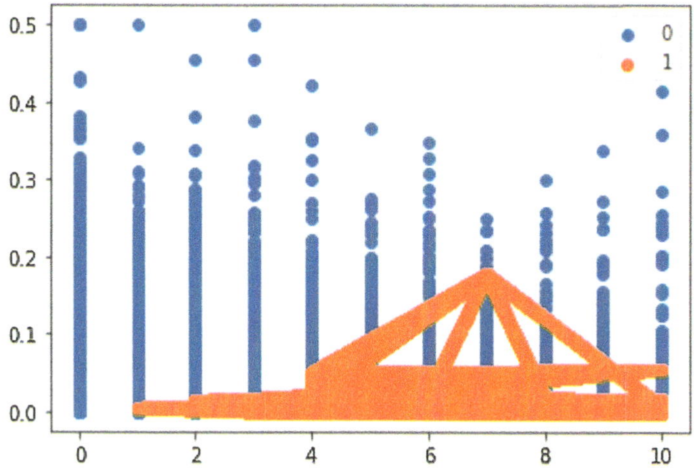

Fig. 4 After oversampling

dataset. Stratified K-folds is performing the sampling in a way the class proportions in individual subsets reflect the proportions in the original dataset. The average of the performance measured for each K-fold is the performance of the cross-validated performance. Each training set of each fold is oversampled by using SMOTE.

4.4.3 SMOTE with Stratified K-Fold Cross-Validator and Random under Sampler Class

Random Under Sampler is a method that is used to under-sampling the majority classes by removing part of the majority classes randomly. In this study, SMOTE and Random Under Sampler are combined to improve the ML performance. The training dataset is firstly under-sampled to have a specific percentage (10% is used in this study) of the majority class. Then, the majority class are under-sampled using Random Under Sampler to have a specific percentage (50% is used in this study) more than the minority class.

5 Analysis and Findings

5.1 Feature Selection

The result of feature selection method shows that non-technical losses are affected by weather. It shows that 40% of top features are related to the summer season (June, July and August) which has a high-temperature degree than other seasons. It also shows from 34 features, 20% of the top 10 features are related to the winter season (December, January, and February). Based on the AccuWeather website, the weathering degree of the summer season of 2020 was 34–45 Celsius in Muscat city. So, there is a relation between the weather condition and anomaly cases in each season.

5.2 SMOTE with Train/Test Split

Based on the result of oversampling of training data got from randomly splitting the dataset, the KNN algorithm has the lowest AUC score which is 61%, and this indicates that the algorithm is slightly unable to distinguish between positive and negative classes due to a lack of discriminating capability. KNN also identifies only 21% of actual positive correctly. Figure 6 shows an approximately straight slope between the true positive rate (TPR) and false-positive rate (FPR). The performance of XgBoost is better than KNN, and it got 78% of the AUC score, and it can identify 40% of the actual positive correctly. Figure 5 shows the ROC graph of XgBoost. The result concludes with ANN outperforming the XgBoost and KNN algorithms as shown in Fig. 7 and got 85% of the AUC score and can identify approximately 73% of the actual positive correctly.

5.3 SMOTE with Stratified K-Fold Cross-Validator

Using stratified K-fold cross-validator with SMOTE is a way to improve the way of splitting the training and testing sets. The average performance measured for each

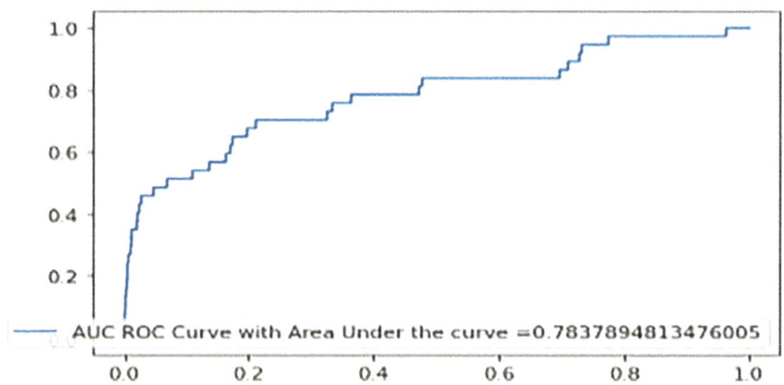

Fig. 5 ROC

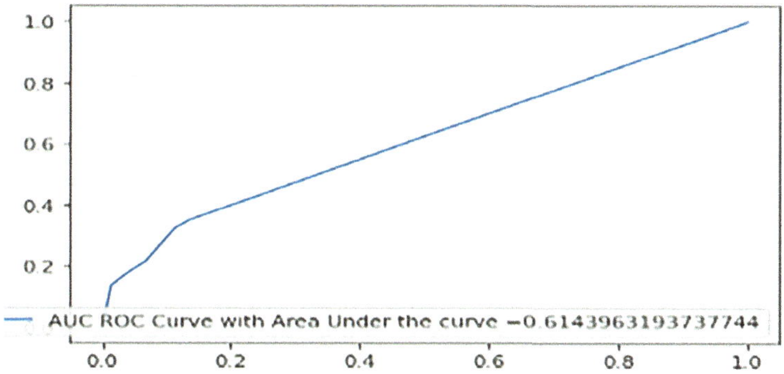

Fig. 6 ROC graph of KNN

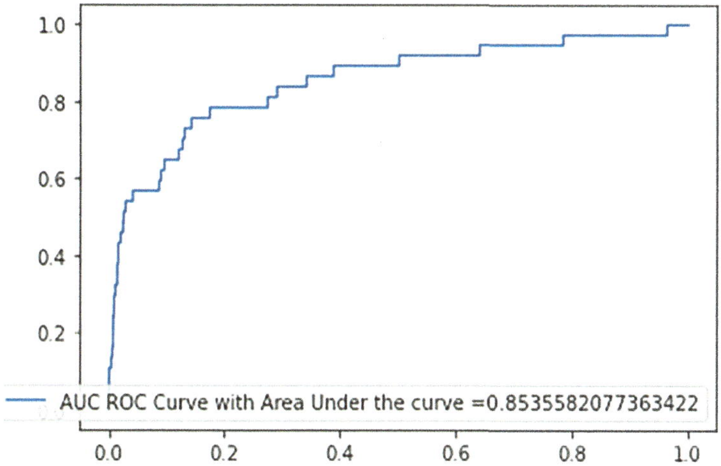

Fig. 7 ROC graph of ANN

K-fold is the performance of the cross-validated performance. It is observed from the results that there is a decreasing trend in AUC scores in all machine learning as shown in Figs. 8, 9, and 10. In addition, there is a decrease in the recall scores. ANN decreased by approximately 3.4%, but it is still able to distinguish between the positive and negative classes, and it decreases approximately 4.2% in recall score. XgBoost is decreasing approximately 6% to be 72% and decreasing 9.2% of recall score to be 31.3%. KNN got 56% of AUC, and the figure shows how the slope is being slightly straight, and it got 16.5% of recall score, and these scores indicate that KNN cannot distinguish between the negative and positive classes and cannot identify the actual positive correctly. The conclusion result shows that ANN still outperformed XgBoost and KNN.

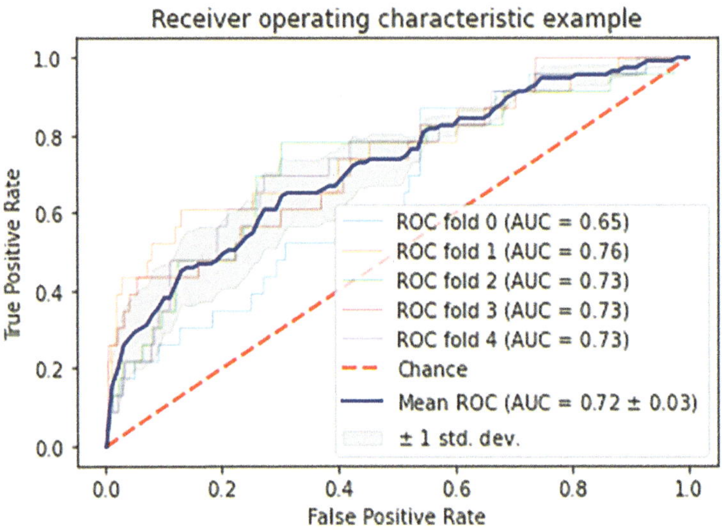

Fig. 8 ROC graph of XgBoost

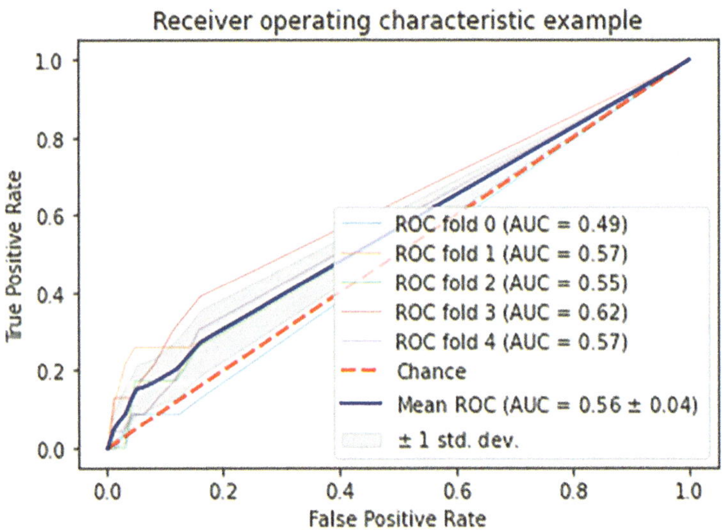

Fig. 9 ROC graph of KNN

5.4 *SMOTE with Stratified K-Fold Cross-Validator and Random Under-Sampler*

Combining the oversampling of the minority class and the under-sampling of the majority class can help to improve the performance of the used machine learning

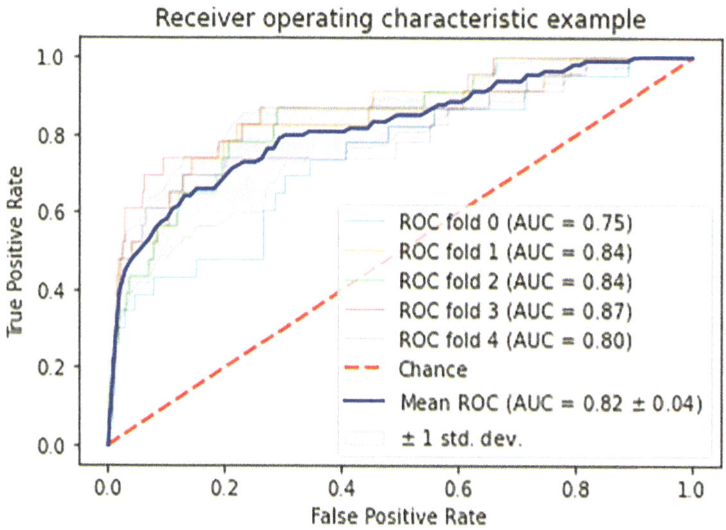

Fig. 10 ROC graph of ANN

algorithms. The results show XgBoost and ANN are getting close scores of AUC to have more than 82% in being able to distinguish between the positive and negative classes. But ANN is 16% better in identifying the actual than XgBoost. The performance of KNN is increased in both identifying the actual positive and in distinguishing between the positive and negative classes. The final result concludes ANN has still outperformed the other two algorithms in both identifying the actual positive correctly and in distinguishing between the positive and negative classes.

5.5 *Using all Features in the Dataset*

Using all features that are extracted from the raw data is a way to check the values of using only the top 10 features to evaluate the performance of the machine learning algorithms that are used in this study. Table 5 summarizes the results of using all extracted features, and Table 6 summarizes the results of using the top 10 features. The results show that using the top 10 features got better results than using all features and the performance of the machine learning algorithms are increased, specially the ANN algorithm showing a high improvement in the performance when using the top 10 features.

The results given in Table 6 show that ANN outperforms the other machine learning algorithms in all methods that are used to evaluate the algorithms. ANN got the best performance when using the SMOTE with the train/test split with 85% AUC score and 70% recall score. XgBoost got the best AUC score in the combination of the oversampling and under-sampling techniques with 83% and got the highest

Table 5 Summary results of using three methods to evaluate the three machine learning algorithms using all features

	SMOTE with train/test split			SMOTE with stratified K-folds			SMOTE and under-sampling with stratified K-folds		
	Accuracy	AUC	Recall	Accuracy	AUC	Recall	Accuracy	AUC	Recall
KNN	0.98	0.58	0.11	0.982	0.53	0.061	0.961	0.614	0.237
XgBoost	0.99	0.79	0.298	0.972	0.71	0.191	0.991	0.802	0.323
ANN	0.73	0.626	0.378	0.609	0.66	0.6	0.946	0.592	0.153

Table 6 Summary results of using three methods to evaluate the three machine learning algorithms using top 10 features only

	SMOTE with train/test split			SMOTE with Stratified K-Folds			SMOTE and undersampling with stratified K-folds		
	Accuracy	AUC	Recall	Accuracy	AUC	Recall	Accuracy	AUC	Recall
KNN	0.932	0.61	0.216	0.935	0.56	0.165	0.939	0.650	0.324
XgBoost	0.977	0.78	0.405	0.933	0.72	0.313	0.978	0.825	0.379
ANN	0.858	0.854	0.729	0.84	0.82	0.687	0.937	0.831	0.535

recall score in using SMOTE with train/test split. KNN got 65% as the highest score of AUC and got 32% as a highest score of recall in the combination of the oversampling and under sampling techniques. Minimizing the number of features plays a good role in increasing the performance of the algorithms.

6 Conclusion

NTL is a primary concern for the electricity companies, and it affects the companies' profit. This research work is implemented and tested using real smart meter data of residential and commercial customers. This study used 2,84,737 normal and anomaly accounts of residential and commercial customers to train the ML algorithms. Thirty-four features were extracted from the raw data, and ten top features were selected to train and test the machine learning models. The dataset was imbalanced because the number of anomalies accounts was less than the normal accounts. So, this study used two ways to solve this issue using SMOTE and Random Under Sampler methods. To detect the NTL, three machine learning algorithms were used, ANN, XgBoost and KNN. The results show that ANN outperformed the two algorithms when using the SMOTE method with the train/test split. It got an 85% of AUC score and 73% of recall score.

References

1. F. de Souza Savian, J.C.M. Siluk, T.B. Garlet, F.M. do Nascimento, J.R. Pinheiro, Z. Vale, Non-technical losses: A systematic contemporary article review. Renew. Sust. Energ. Rev. **147**, 3 (2021)
2. M.S. Saeed, M.W. Mustafa, U.U. Sheikh, T.A. Jumani, I. Khan, S. Atawneh, N. Hamadneh, An efficient boosted C5.0 decision-tree-based classification approach for detecting non-technical losses in power utilities. Energies **13**(12), 1–19 (2020)
3. M. Buzau, S. Member, J.T. Aguilera, P. Cruz-romero, Detection of non-technical losses using smart meter data and supervised learning. IEEE Trans. Smart Grid **10**(3), 2661–2670 (2018)
4. J.A. Dominguez, A. Rivera, K. Botina, G.A. Perdomo, O. Montoya, J. Campillo, E. Delahoz, Data-driven framework for the detection of non-technical losses in distribution grids, in *2020 IX International Congress of Mechatronics Engineering and Automation (CIIMA)*, (Cartagena, 2020), pp. 1–6, IEEE
5. B.C. Puig, J. Carmona, Bridging the gap between energy consumption and distribution through non-technical loss detection. Energies **12**(9), pp. 1, 7, 11–13 (2019)
6. G.M. Messinis, N.D. Hatziargyriou, Review of non-technical loss detection methods. Electr. Power Syst. Res. **158**, 250–266 (2018)
7. M.A. Uparela, R.D. Gonzalez, J.R. Jimenez, C.G. Quintero, Intelligent system for non-technical losses management in residential users of the electricity sector. Ing. Investig. **38**(2), 52–60 (2018)
8. K.M. Ghori, R.A. Abbasi, M. Awais, M. Imran, A.T.A. Ullah, L. Szathmary, Performance analysis of different types of machine learning classifiers for non-technical loss detection. IEEE Access **8**, 16033–16048 (2020)

9. R. Razavi, A. Gharipour, M. Fleury, I. Justice, A practical feature-engineering framework for electricity theft detection in smart grids. Appl. Energy **238**, 481–494 (2019)
10. A. Abu-Siada, Y. Yao, Detection of electricity theft behavior based on technique and random Forest classifier. Energies **13**(8), pp. 3, 5, 13–16 (2020)
11. R. Punmiya, S. Choe, Energy theft detection using gradient boosting theft detector with feature engineering-based preprocessing. IEEE Trans. Smart Grid. **10**(2), 2326–2329 (2019)
12. P. Chandrasekar, K. Qian, The impact of data preprocessing on the performance of Naïve Bayes, in *IEEE 40thAnnual Computer Software and Applications Conference (COMPSAC)*, (Atlanta, 2016), pp. 40–41, IEEE
13. DeepAI, Feature extraction [Online]. Available: https://deepai.org/machine-learning-glossary-and-terms/feature-extraction

Part II
Bigdata and Security

Data Hiding in Binary Images for Secret and Secure Communication Using Decision Tree

Gyankamal Chhajed and Bindu Garg

Abstract In today's digital era, secret communication through digital media has become difficult due to advances in technology. Many methods are used to maintain secrecy like password protection, cryptography, etc. The drawback of applying these techniques is that it alerts about the presence of secret information within media. So information hiding in the multimedia medium is the most powerful mean for secret communication. Binary images are frequently transferred and so can be utilized as a carrier for hiding secret information. But hiding in binary images is challenging as changes caused due to hiding are very easily noticeable and hence limit hiding capacity. Image distortion, information hiding capacity, security, and imperceptibility are the common factors to evaluate the performance of any data hiding techniques. In existing techniques, either the data hiding capacity or visual distortion has been taken care of. This paper proposes a novel approach for hiding data in binary images which focuses on minimization in visual distortion providing good data hiding capacity and security to hidden data. To achieve this, the original image pattern is utilized in the best possible way and the message is hidden distributing over the image in the encrypted form. In this technique, the image and secret message are divided into parts where image partitions are further divided into 3×3 pixel blocks. These image partitions and secret message parts are provided to the decision tree for deciding the best suitable image partition for hiding secret message parts. The decision tree takes decisions based on the type of match of LRUD pattern of 3×3 pixel block and the message part pattern to be hidden. The receiver can retrieve the data without the original image with the help of

G. Chhajed
BVDUCOE, Pune, Maharastra, India

Vidya Pratishthan's Kamalnayan Bajaj Institute of Engineering and Technology, Baramati, Maharastra, India

B. Garg (✉)
Bharti Vidyapeeth Deemed to be University College of Engineering, Pune, Maharastra, India
e-mail: brgarg@bvucoep.edu.in

A. Haldorai et al. (eds.), *4th EAI International Conference on Big Data Innovation for Sustainable Cognitive Computing*, EAI/Springer Innovations in Communication and Computing, https://doi.org/10.1007/978-3-031-07654-1_5

indicator blocks which are prepared at the time of embedding. The results show that the proposed approach provides good hiding capacity minimizing distortion caused due to data hiding.

Keywords Binary image · Information security · Steganography · Watermarking · Decision tree

1 Introduction

Today, digital media are getting more and more popular. The binary document is also digitized as an image in many applications including legal documents, digital books, maps, and architectural and electronic drawing. In addition, more important documents such as fax, insurance, and personal documents are digitized and stored. Nowadays, people prefer images of information to avoid time-consuming writing, preserving, or transferring as it is possible to have it with just a few clicks. In such an open world of information, exchanging messages secretly, safely, and securely is an important topic of research. Media files serve two purposes. One is, to provide information through visible content and another is, to act as a carrier for invisible secret information for covered communication. Originality and authenticity matters when its visible content is important whereas the secrecy and safety of invisible content is important when used as carrier. There are many ways or methods used before for hiding data in images which usually distorts images. Most of the spatial domain data hiding techniques use image blocks. Block size, criteria for identifying the suitability of block to carry information, number of pixel modification possible, and location of the pixel in a block make all techniques unique. As per the study, generally, 1 or 2 bits are hidden in a 3×3 pixel block. To judge the performance of the technique, hiding capacity and visual distortion are the main concern. Techniques providing a high data hiding capacity with less visual distortion are considered as better performing techniques. But satisfying this is very difficult in binary images due to the use of only two shades. At the same time, these factors are inversely correlated. If it is tried to increase data hiding capacity, it imposes more visual distortion, and if the visual distortion is taken care of, hiding capacity gets limited. So balancing these two factors is a topic of research.

2 Literature Survey

Image encryption and support vector machine (SVM) is used for data hiding, extraction, and image recovery [1]. It uses image blocks, and in each block, the one-bit message can be embedded. In this scheme, encryption keys have been used at three levels, and only the authorized receiver can get the hidden message. A support vector machine (SVM) model is also trained for extraction. A small block-based technique is proposed in [2] in which information is hidden adapting image block pattern. The technique claims the highest data hiding rate. It applied the principle of Webb's law and used blind data extraction. Data hiding and extraction algorithms have very low computational complexity. In [3], a binary image steganalysis scheme is proposed which checks the effect of embedding and identifies that the data is hidden. The one-shape pattern is considered, and distortion is calculated for four classes of patterns. It used images to constitute a 32-dimensional steganographic feature set. Further, a two-dimensional feature set that only involves the distribution of two patterns is computed. They are then combined with the soft-margin support vector machine (SVM) to construct steganalysis. The proposed scheme presents a significant performance in detecting state-of-the-art data hiding schemes. In [4], the combination theory is applied to reduce visual distortion. A special matrix is used which increases hiding capacity reducing the visual artifacts. In this technique, one pixel is changed at most for each block, and $[\log 2\ (m \times n + 1)]$ bits are hidden into a block of size $m \times n$. The scheme yields capability of a higher payload embedding maintaining noise. Block classification and local complexity calculation are used in [5]. In this scheme, at most, only two pixels are modified to embed three bits of data into each selected block. The complex region of the image is used to hide the secret data so that it is difficult to detect whether data is hidden or not. This also avoids grabbing attention from attackers. The texture property of binary images is considered in [6] and proposed an efficient binary image data hiding with similarity matching method based on texture. A new information security method is proposed in [7]. The proposed technique targets a more unnoticeable and secure transmission of data that is secretly hidden compared to the generally used LSB algorithm. A novel scheme for authentication of a binary image is proposed in [8] that has a small distortion in the cover image. Hamming codes-based scheme is used in the data-embedding algorithm and is proposed in [9] where the ELSSM (Edge Line Segment Similarity Measure) algorithm is used for selecting and flipping pixels. A new multi-class steganalysis method was proposed in [10]. The proposed method is capable of identifying the technique used for hiding data. The data hiding in binary text documents is proposed in [11] where the 8-connected boundary of character is used for embedding. A five-pixel boundary with a fixed set of pairs of patterns is utilized to embed data by adding or deleting the center foreground pixel in a pair. An image quality maintaining scheme is proposed in [12] after data hiding. The imperceptibility after embedding is identified by comparing the modified bit and its adjacent bits. In [13] discrete cosine transform (DCT) domain is used by checking for watermark embedding possibility in the binary images. The method in [14–16] uses relationships within the block to identify "flippable" pixels which cause minimum visual distortion. Visual alteration detection and accordingly deciding hiding data in such a way that visual quality is maintained for all types of binary images is proposed in [17]. A variety of proposed schemes also presented which locates updated regions. The connectivity of pixels in a neighborhood preserving method is proposed in [18]. In a 3×3 moving window, the "flippability" is identified centered at a pixel. In [19], a blind binary image authentication scheme uses a two-layer where the first layer is used for the authentication and the second layer locates the tampering locations. A scheme that uses the morphological transform domain is proposed in [20, 21]. proposed reversible data hiding scheme where run-length (RL) histogram is monitored and modified to embed data. The sequence of RL couples was used to hide data that are formed by black RL and white RL. Review and analysis of exiting watermarking and steganography techniques are presented in [22] considering image processing in spatial and transform domain. The regression analysis model was used to accomplish the efficient de-fuzzification operation [23]. An ordered weighted aggregation (OWA) proposes in [24] for fuzzy time series and further designed forecasting model signifying efficacy of the proposed concept. RSA, triple DES in [25], and homomorphic encryption techniques are discussed in [26]. An adaptive overlapping embedding is proposed, which uses the optimum pair of the local context for embedding with a three-round embedding mechanism [27]. A spatial domain-based scheme proposed a novel flipping distortion measurement that considers both HVS and statistics [28]. A fragile data hiding technique proposed for image authentication which identifies the alteration in the image if carried out [29]. A novel pixel-wise data hiding method for the hybrid authentication scheme is proposed. Its core engine invariantly selects a large number of good DCPLs in both embedding and extraction [30]. A general methodology is proposed for embedding while minimizing an arbitrary additive distortion function with performance near the theoretical bound [31]. A new framework is proposed for designing distortion functions of joint photographic experts group (JPEG) steganography with uncompressed side images [32]. Covert communication over different file formats is proposed applying encryption and decryption [33].A four-way scan method is used, and the best scan method is used to substitute information to be hidden which provides minimum distortion [34]. The edge portion of the host binary image is used which locates the best changeable pixels in a block by changing the distance matrix dynamically and computing its changeable score by weighting mechanism [35]. The development and analysis of the digital watermark (DWM) are proposed which enhances the robustness [36]. A new steganography scheme using a reference image is proposed with two variants; secret text is written/printed on a white canvas, and then that image containing the secret text is converted into a binary (BW) image [37]. An approach for image steganography with high performance proposed in which an extreme learning machine (ELM) algorithm is modified to create a supervised mathematical model [38]. The data-embedding algorithm is based on hamming codes, which flips only a small number of pixels to minimize visual distortion [39]. Fuzzy logic interval-based partitioning is proposed in different intervals [40]. An RDH scheme by flipping pattern pairs with opposite center pixels (PPOCPs) in binary images is proposed [41]. A new reversible data hiding scheme in an encrypted binary image is proposed by shared pixel prediction and halving compression [42].

3 Proposed Model

It uses image blocks, and in each block, the one-bit message can be embedded. The proposed model provides a novel approach to hide data in a 3 × 3 pixel block. It focuses on the selection of suitable blocks for hiding in such a way that it will introduce less distortion providing high data hiding capacity. The novelty of this model is lying in the utilization of a trending decision tree for selecting suitable blocks in images for information hiding.

3.1 Image Partitioning and Message Division and Grouping

As shown in Fig. 1(a), the image is divided into blocks of size 3 × 3 pixels. These 3 × 3 pixels blocks are grouped in such a way that the whole image looks to be divided into four equal horizontal parts nominated as B1, B2, B3, and B4. Data to be hidden called message is represented as a bitstream. Encryption is applied to secure this stream using a key known to sender and receiver. This encrypted bitstream is partitioned into four parts say M1, M2, M3, and M4 where each part has groups of 4 bits as shown in Fig. 1b. The 64 bits are divided into four parts of 16 bits where each part is further divided into groups of 4 bits.

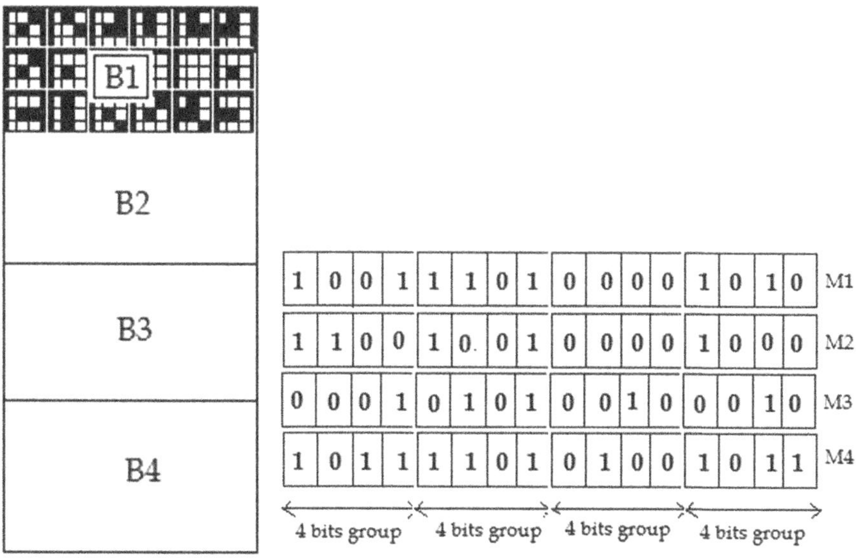

(a) (b)

Fig. 1 (**a**) Image partitioning, (**b**) message division and grouping

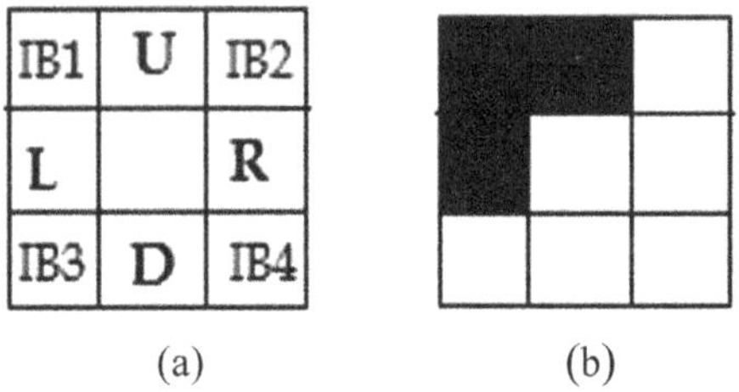

Fig. 2 (**a**) LRUD pattern representation for 3 × 3 pixel block (**b**) e.g. LRUD: 1010, Indicator bits 1000

3.2 Block Processing

In this model, a 3 × 3 pixel block is used to hide information. Each 3 × 3 image block has the left, right, up, down LRUD pixel pattern as shown in Fig. 2(a). This pattern is utilized to hide the 4-bit message group. To hide these 4 bits, the suitability of the block needs to be judged. For this judgment, the decision tree is utilized. The decision of the decision tree depends on the corner bits cumulative pattern. The corner pixels of the 3 × 3 block indicated as IB1, IB2, IB3, and IB4 cumulative patterns justify the existence of hidden message bits. If the LRUD pattern is matching with supplied 4 bits and the corner 4 pixels have black and white pixel count as 4,0 or 0,4 or 2,2, then that indicates the block is an indicator block and is suitable for embedding without flipping any pixel. If only the LRUD pattern matches and does not satisfy the criteria for indicator block, then it is said it's a non-indicator block. As shown in Fig. 2(b), the indicator bit pattern is "1000" which has a black and white pixel count of 1, 3 which is a non-indicator block. So if the LRUD pattern is "1010" and if the decision tree selects the image part having this block pattern, then one white pixel needs so that count of black and white pixel count will become 2,2 and will be an indicator block.

3.3 Decision Tree for Selection of Block for Data Hiding

The proposed model focuses on the distribution of hidden data over the image to minimize visual artifacts as shown in Fig. 3. The utilization of a decision tree for this purpose provides novelty to this technique. The decision tree identifies the most suitable image part which supports minimization of distortion satisfying the best possible distribution of hidden data over the image. The selection of the suitable block depends not only on the LRUD pattern and indicator bit pattern but also on the image partition in which the block is located and the sequence of the 4-bit message group. For distribution of message in all parts, it is expected B1 should accommodate M1 and similarly other respectively. But in reality, it may be the case that all indicator blocks are not in B1 for M1 and other pairs as well. So

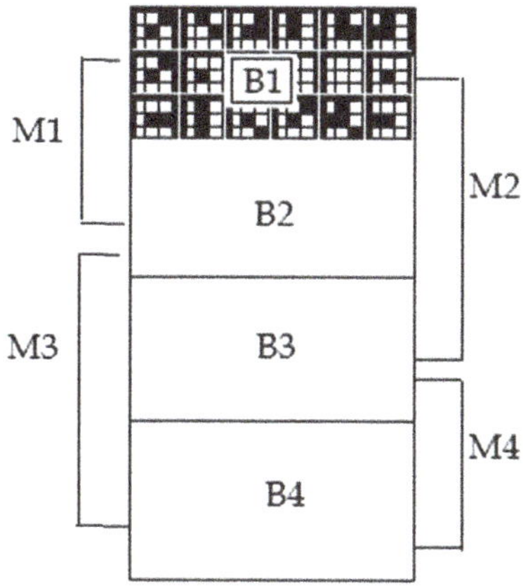

Fig. 3 Expected message block embedding in respective image partition

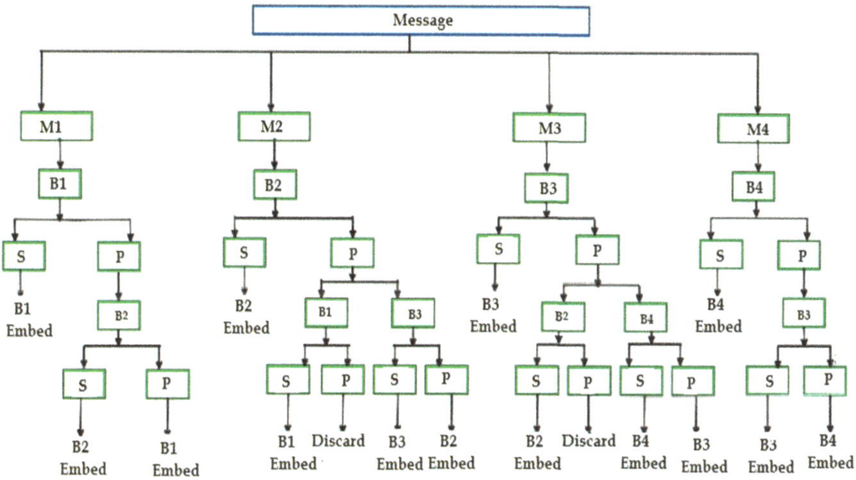

Fig. 4 Decision tree

predecessor and successor blocks are also considered. The decision tree shown in Fig. 4 is used to get a decision about which block is most suitable for embedding supplied message block. If the decision is received through fully matching criteria, i.e., indicator block denoted as "S," 4-bit data is embedded without any change in the 3×3 block If selection is through the partial matching branch, then the corner pixel is flipped to satisfy the indicator block criteria. In the same way, all message blocks are embedded, and the image is sent to the receiver.

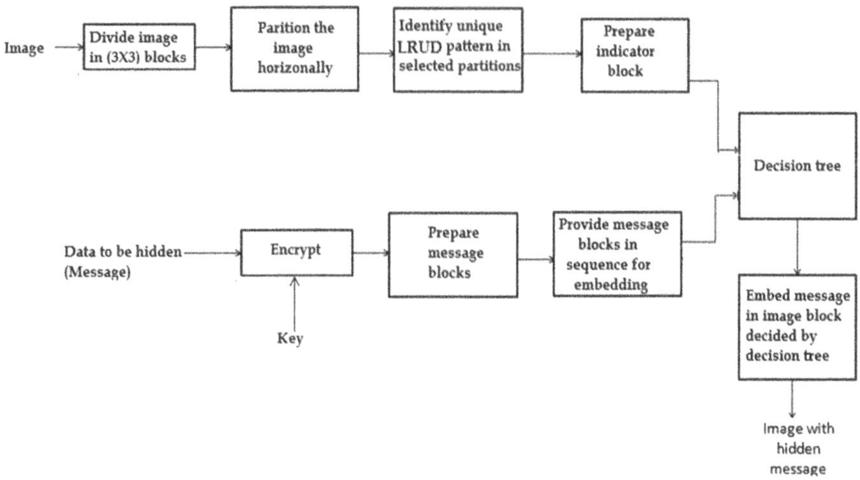

Fig. 5 Data embedding process

3.4 *Embedding Process*

The data embedding process is shown in Fig. 5. Very first, the image is partitioned into blocks of size 3 × 3 pixels. Then the image is partitioned into image parts horizontally as shown in Fig. 1. Then the unique LRUD patterns are identified to represent which LRUD pattern is carrying message bits and prepared as indicator block. Then data to be hidden, called message, is represented in bitstream and encrypted using a key. Then message blocks are formed where each message block has a set of 4 bits message groups. The 4 bits message groups are provided to the decision tree to get the image part which most suitable for hiding those message group. The 4-bit message group is embedded in the image block decided by the decision tree. Before embedding the indicator, bit patterns are checked. If it is a non-indicator block, then one pixel is flipped in a block, depending on the count of the black and white pixel in a block.

3.5 *Extraction Process*

The extraction process is shown in Fig. 6. First and foremost, the image is divided into blocks of size 3 × 3 pixels. Image is partitioned into image partitions horizontally distributing 3 × 3 pixels blocks equally. To extract an exact number of hidden bits, first, the length information is collected. Then through the indicator blocks, 4 bits are extracted from the LRUD pattern till all message bits are extracted. The extracted message bits are combined and decrypted with the shared key among sender and receiver. The original message is obtained from decrypted bits.

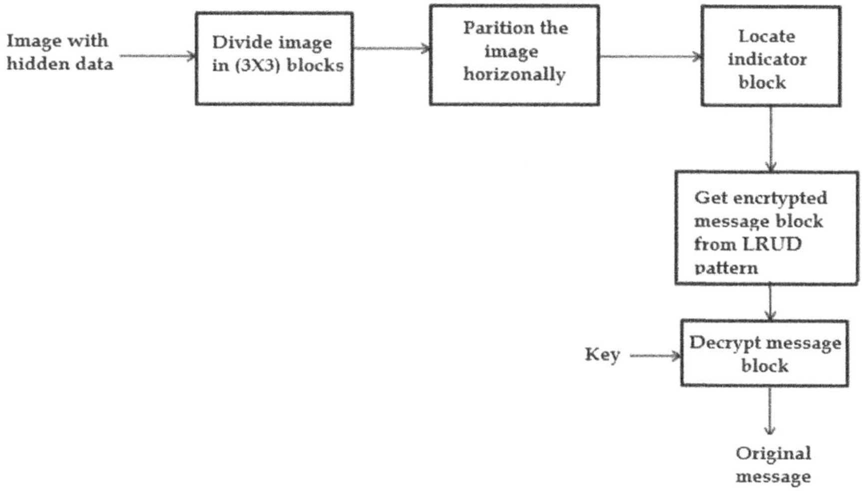

Fig. 6 Data extraction process

4 Algorithms

The decision tree is an algorithm to get decisions based on existing different situations. The proposed work is built considering all possible situations and is capable of providing decisions for the supplied situation. In the proposed system, it decides which image partition block is suitable for the supplied message 4 bits to avoid unnecessary distortion due to the flipping of pixels. As shown in Fig. 3, "S" indicates a "suitable" image partition block for embedding for supplied set of message 4 bits. This is based on a match of LRUD pattern of a 3 × 3 image block and indicator bits. "P" indicates partial matching where LRUD patterns are matching, but indicator bits are not satisfying the criteria.

Algorithm for Embedding
Input: Image, Message
Output: Image with a hidden message

Step 1. Divide the image into four parts as B1, B2, B3, and B4 where each block is having equally distributed 3 × 3 pixel blocks.
Step 2. Get the data to be hidden called message, and convert it into a binary stream.
Step 3. Get the length of the message bitstream, and append this length to the message bitstream.
Step 4. Represent above bitstream in multiples of 4 bits.
Step 5. Encrypts data with a shared key.

Step 6. Get the encrypted information to be hidden called messag,e and divide it into four parts as M1, M2, M3, and M4 where every message block has a group of 4 bits.

Step 7. Pass group of 4 bits, one by one to decision tree for all message parts in sequential order.

Step 8. Get decision about the selection of block for embedding each group

Step 9. If it is selected through suitable, indicated as 'S' path, then

Embed in that block without any flipping, and it is indicator block.
Else if selected through partially suitable, indicated as "P" path, then
Flip one pixel so that it will become an indicator block.

Step 10. Check the first occurrence of the unique LRUD pattern block, and if matching with message 4-bit pattern, convert it into indicator block.

Decision Tree

Input: Image blocks, Message parts

Output: Decision of selection of blocks for embedding

Step 1. Get the partitioned image in four parts with equally distributed 3×3 pixel blocks.

Step 2. Get message bits grouped in 4 bits.

Step 3. Check how M1 is matching with the LRUD pattern of 3×3 blocks in B1
If M1 is fully matching with LRUD pattern of 3×3 block in B1, then
B1 is selected.

Else check the next block B2 if matching partially in B1.
If M1 is fully matching with LRUD pattern of 3×3 block in B2, then
B2 is selected.
Else B1 is selected.

Step 4. Check how the LRUD pattern of 3×3 blocks in B2 is matching the pattern of M2

If M2 is fully matching with LRUD pattern of the 3×3 block in B2, then
B2 is selected.
Else check the blocks B1and B3.
If M2 is fully matching with LRUD pattern of 3×3 block in
B1or B3, then respective block selected.
Else B2 is selected.

Step 5. Similarly selected blocks are provided for other message blocks M3 and M4.

Algorithm for Extraction
Input: Image with hidden data
Output: Hidden data

Step 1. Divide the image into four parts as B1, B2, B3, and B4 where each block is having equally distributed 3×3 pixel blocks
Step 2. Check the indicator bits for the first occurrence of unique LRUD pattern block, and get the patterns that are the part of the hidden message.
Step 3. Locate indicator block, and extract hidden message in groups of 4 bits from LRUD pattern.
Step 4. Decrypt extracted encrypted message, and get the length of the message.

Extract message based on indicator bits from all blocks.

Step 5. Combine extracted bits and get hidden message bitstream.

5 Result Analysis and Performance Evaluation

Experiments on standard images and other sample images are conducted using NetBeans IDE 8.2, JAVA 8. The mean squared error (MSE) and peak signal-to-noise ratio (PSNR) are standard measurement generally used as metrics for performance evaluation of data hiding techniques. The high value of PSNR indicates less distortion after hiding data in image. If the cover image is C of size $M \times N$ and the image after data hiding is S of size $M \times N$, then C and S image will have pixel value (x,y) from 0 to M-1and 0 to N-1, respectively. The PSNR is then calculated as follows:

$$\mathrm{MSE} = \frac{1}{\mathrm{MN}} \sum\nolimits_{x=0}^{M-1} \sum\nolimits_{y=0}^{N-1} (C(x, y) - S(x, y))^2 \tag{1}$$

$$\mathrm{PSNR} = 10.\log_{10}\left[\frac{\mathrm{MAX}^2}{\mathrm{MSE}}\right] \tag{2}$$

The technique is tested with set of standard images and sample figures are shown in Fig. 7 before and after hiding data and results are analyzed in Table 1 and Figs. 8, 9, and 10. The PSNR and MSE are used to check the performance of the proposed technique. A textual message "success" represented in 56 bits encrypted message is embedded in the images.

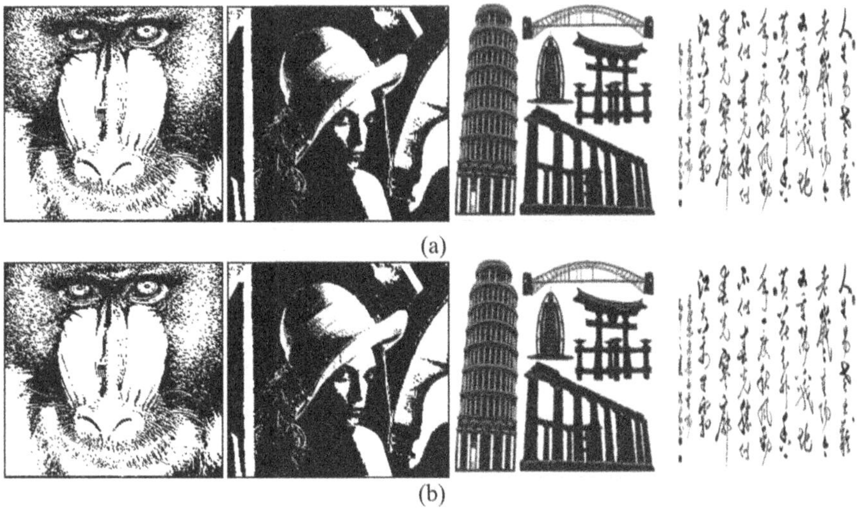

Fig. 7 Sample images for testing (**a**) before data hiding (**b**) after data hiding

6 Discussion

From the obtained results, it is observed that flipped pixel count range is 15–24 for hiding the same data of size 56 bits. The number of flipped pixels is dependent on the size of the image, LRUD pattern in the image block, and data block bit pattern. The visual distortion is dependent on the matching of LRUD pattern available in image blocks with the 4-bit message block, more the match less will be the distortion. The decision tree takes care of selecting a suitable block that causes minimum distortion. In many block-based techniques, 1 or 2 bits are embedded in the 3×3 pixel block where in this method, 4 bits are hidden in 3×3 block. To hide 4 bits at maximum, 1 bit is changed if required. This proves proposed technique provides good hiding capacity and also since original pattern of image is utilized at maximum, it results in minimization of distortion. One more benefit is that hiding of message bits is distributed among image partitions; it is difficult to locate hidden information bits.

7 Conclusion and Future Scope

The proposed data hiding technique provides a novel approach of using decision tree for hiding data in binary images. It collectively provides the benefits of high hiding capacity, security of hidden data, and minimized distortion. It uses a decision tree which selects suitable block for data hiding. In this technique, a specific pattern of 4 bits of 3×3 pixel block called as LRUD pattern is matched with a block of 4-bit

Table 1 Result analysis table with sample images

Sr. No.	Image	Image size	Total pixels	Image DPI	Message size (no of bits)	Flipped pixel count (no. in bits)	MSE 10-3	PSNR
1	Baboon	200 × 200	40,000	96	56	15	5.63	32.04
2	Lena	299 × 295	88,205	96	56	22	5.49	32.15
3	Image 1	256 × 256	65,536	96	56	17	4.41	33.10
4	Image2	300 × 298	83,400	96	56	24	6.91	31.14
5	Image3	297 × 172	51,084	96	56	16	5.01	32.54

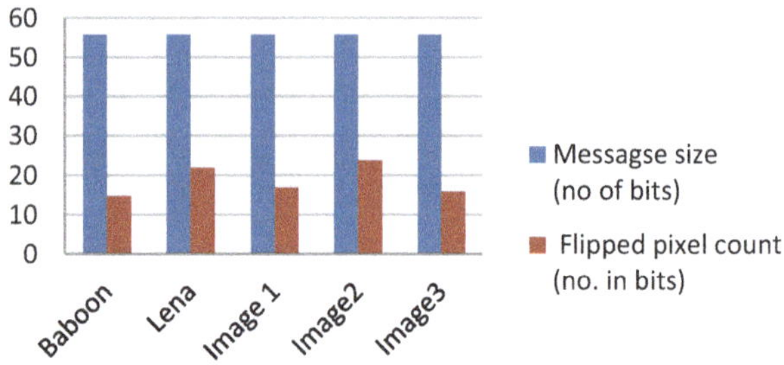

Fig. 8 Comparison of size of message embedded versus flipped pixels

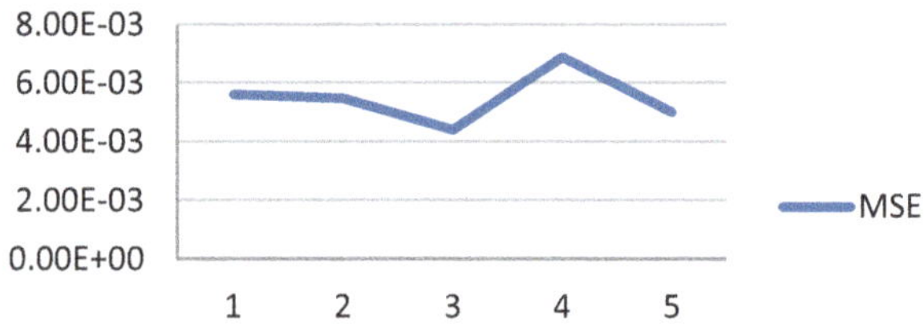

Fig. 9 Performance analysis using MSE

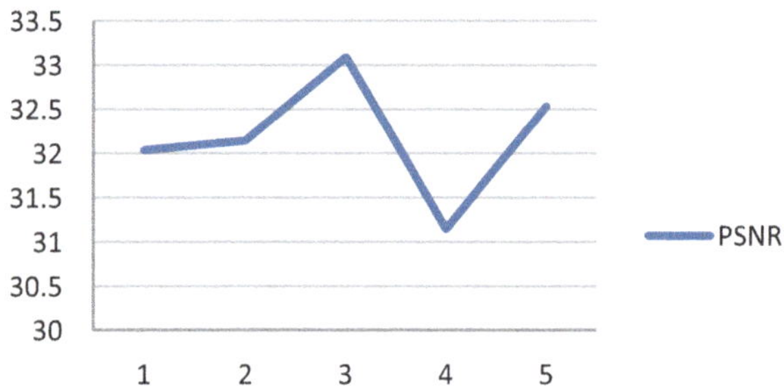

Fig. 10 Performance analysis using PSNR

encrypted data to be hidden and accordingly decided whether the 3×3 pixel block is suitable for data hiding or not. The data is hidden after the encryption applied in the scattered form so that interpreting and locating data is difficult for unauthorized users. Since the message block is matched with LRUD patterns of image blocks and indicator bits, it minimizes distortion as the only pixel at maximum required to change to embed 4 bits in a 3×3 block if it is not suitable to be utilized as it is. So it hides 4 bit in single block as opposed to other block-based techniques which hides 1-bit or 2-bit information. The experimental result shows that the proposed scheme significantly uses existing image patterns flipping very few pixels to hide whole message which proves that technique provides high hiding capacity with less distortion. In future, the rotation, mirroring of LRUD pattern available with image block can be utilized for matching the 4-bit message block to increase the possibility of using same image block for hiding various patterns.

References

1. V.M. Manikandan, V. Masilamani, Reversible data hiding scheme during encryption using machine learning.", Elsevier Science Direct. Procedia Comput. Sci. **133**, 348–356 (2018)
2. W. Ding, Y. Wang, Data hiding in binary image with high payload., Springer. Arab. J. Sci. Eng. **43**, 7737–7745 (2018). https://doi.org/10.1007/s13369-018-3130-5
3. B. Feng, J. Weng, Stegnalysis of content adaptive binary image data hiding. J. Vis. Commun. Image Represent. **46**, 119–127. Elsevier ScienceDirect 2017 (2017). https://doi.org/10.1016/j.jvcir.2017.01.008
4. W. Nan-I, M.-S. Hwang, Development of data hiding schema based on combination theory for lowering the visual noise in binary image. Elsevier ScienceDirect (2017). https://doi.org/10.1016/j.displa.2017.07.009
5. W. Han-Zhou, W. Hong-Xia, S. Yun-Quing, Separable reversible data hiding for encrypted palette images with color partitioning and flipping verification. IEEE (2016). https://doi.org/10.1109/TCSVT.2016.2556585
6. M. Honey Mol, P.I. Reji, A secure binary data hiding and comparison technique. Int. J. Sci. Eng. Res. **7**(7), 810 (2016). ISSN 2229-5518.
7. T.-S. Nguyen, C.-C. Chang, *High capacity data hiding for binary image based on block classification* (Springer, 2015). https://doi.org/10.1007/s11042-015-2768-1
8. S. Udhayavene, A.T. Dev, New data hiding technique in encrypted image: DKL algorithm.", Elsevier ScienceDirect. Procedia Comput. Sci. **54**, 790–798 (2015)
9. Y. Lee, H. Kim, A new data hiding scheme for binary image authentication with small image distortion." Springer. Procedia Comput. Sci. Informat. Sci. **179**(2009), 3866–3884 (2011)
10. K.L. Chiew, J. Pieprzyk, Binary image steganographic techniques classification based on multi-class steganalysis. Springer ISPEC 2010, LNCS 6047, pp. 341–358, Springer-Verlag Berlin Heidelberg 2010
11. Q. Mei, E.K. Wong, N. Memon, Data hiding in binary text document. Proc. SPIE **4314**, 369–375 (2001)
12. Y.C. Tseng, H.-K. Pan, Data hiding in 2-color images. IEEE Trans. Computer. **51**(7), 873–878 (2002)
13. H. Lu, X. Shi, Y.Q. Shi, A.C. Kot, L. Chen, Watermark embedding in DC components of DCT for binary images, in *Proc., IEEE Workshop on Multimedia Signal Processing*, (2002), pp. 300–303
14. M. Wu, B. Liu, Data hiding in binary images for authentication and annotation. IEEE Trans. Multimedia **6**(4), 528–538 (2004)
15. G.J. Chhajed, V.S. Inamdar, V.Z. Attar, Steganography in black and white picture images, in *2008 Congress on Image and Signal Processing, CISP 2008, IEEE Computer Society*. 978-0-7695-3119-9/08, pp. 141–144. https://doi.org/10.1109/CISP.2008.626
16. G.J. Chhajed, S.A. Shinde, Efficient embedding in B&W picture images, in *2nd IEEE International Conference ICIME 2010*, (2010, 978-1-4244-5265-1/10), pp. 525–528
17. H.Y. Kim, R.L. de Queiroz, Alteration-locating authentication watermarking for binary images. Proceedings of International Workshop Digital Watermarking **2005**, 125–136 (2004., Springer-Verlag Berlin Heidelberg)
18. H. Yang, A.C. Kot, Pattern-based date hiding for binary images authentication by connectivity-preserving. IEEE Trans. Multimedia **9**(3), 475–486 (2007)
19. H. Yang, A.C. Kot, Binary image authentication with tampering localization by embedding cryptographic signature and block identifier. IEEE Sig. Process. Lett. **13**(12), 741–744 (2006)
20. H. Yang, A.C. Kot, S. Rahardja, Orthogonal data embedding for binary images in morphological transform domain-a high capacity approach. IEEE Transact. Multimedia **10**(3), 339–351 (2008)
21. G. Xuan, Y.Q. Shi, P. Chai, X. Tong, J. Teng, J. Li, Reversible binary image data hiding by run-length histogram modification. International Conference on Pattern Recognition, 1651–1653 (2008) 978-1-4244-2175, IEEE.

22. C.G.J. Krupali Deshmukh, T. Kulkarni, Review on binary image stegnography and watermarking. Int. J. Comput. Sci. Eng. **3**(11), 3545–3651 (2011)., ISSN: 0975-3397
23. B. Garg, R. Garg, Enhanced accuracy of fuzzy time series model using ordered weighted aggregation." Elsevier. Appl. Soft Comput. **48**, 265–280 (2016)
24. B. Garg, S. Aggarwal, J. Sokhal, Crop yield forecasting using fuzzy logic and regression model. Comput. Electr. Eng. **67**, 383–403 (2018)
25. R. Jain, S. Madan, B. Garg, Implementation and comparison of RSA and triple DES algorithm for encryption and encryption in cloud environment. Int. J. Appl. Eng. Res. (SCOPUS Indexed) **10**, 11957–11971 (2015). ISSN 0973-4562.
26. R. Jain, S. Madan, B. Garg, E-voting system using homomorphic encryption in a cloud based environment. Int. J. Adv. Comput. Sci. Applicat. (Web of Science) **11**, 59–68 (2017). ESCI index
27. K. Dong et al., Reversible data hiding for binary images based on adaptive overlapping pattern. EURASIP J. Inf. Secur. **2020**, 11 (2020). https://doi.org/10.1186/s13635-020-00107
28. B. Feng, W. Lu, W. Sun, Secure binary image steganography based on minimizing the distortion on the texture. IEEE Trans. Inf. Foren. Secur. **10**(2), 243–255 (2015). https://doi.org/10.1109/TIFS.2014.2368364
29. M. Guo, H. Zhang, High capacity data hiding for binary image authentication, in *Proceedings of International Conferences Pattern Recognition*, (2010), pp. 1441–1444. https://doi.org/10.1109/ICPR.2010.356
30. H. Cao, A.C. Kot, On establishing edge adaptive grid for bilevel image data hiding. IEEE Trans. Inf. Foren. Secur. **8**(9), 1508–1518 (2013)
31. T. Filler, J. Judas, J.J. Fredric, Minimizing additive distortion in steganography using syndrome-trellis codes. IEEE Trans. Inf. Foren. Secur. **6**(3), 920–935 (2011)
32. T. Pevn, T. Filler, P. Bas, Using high- dimensional image models to perform highly undetectable steganography, in *Information Hiding (Lecture Notes in Computer Science)*, ed. by R. Bhme, P. W. L. Fong, R. Safavi-Naini, vol. 6387, (Springer-Verlag, New York, 2010), pp. 161–177
33. S.N. Bal, M.R. Nayak, S.K. Sarkar, On the implementation of a secured watermarking mechanism based on matching. J. King Saud Univ. Comput. Inf. Sci. (2018). https://doi.org/10.1016/j.jksuci.2018.04.006
34. G.J. Chhajed, B.R. Garg, Novel approach to data hiding in binary images minimizing distortion, in *Computational vision and bio-inspired computing. Advances in intelligent systems and computing*, vol. 1318, (Springer, Singapore, 2021), pp. 587–598. https://doi.org/10.1007/978-981-33-6862-0_46
35. J. Chen, L. Wei, Y. Fang, X. Liu, Y. Yeung, Y. Xue, Binary image steganalysis based on local texture pattern. Accepted manuscript in J. Vis. Commun. Image R. 1–20 (2018), https://doi.org/10.1016/j.jvcir.2018.06.004
36. T.H. Liu, L.-W. Chang, An adaptive data hiding technique for binary images, in *17th International Conference on Pattern Recognition (ICPR'04)*, (2004). https://doi.org/10.1109/ICPR.2004.1333901
37. G. Maji, S. Mandal, Secure and robust image steganography using a reference image as key. Int. J. Innov. Technol. Explor. Eng. **8**(7), 2828–2837 (2019) ISSN: 2278-3075
38. H.A. Atee, R. Ahmad, N.M. Noor, A.M.S. Rahma, Y. Aljeroudi, Extreme learning machine based optimal embedding location finder for image steganography. PLoS One **12**(2), e0170329 (2017). https://doi.org/10.1371/journal.pone.0170329
39. K.L. Chiew, J. Pieprzyk, Binary image steganographic techniques classification based on multi-class steganalysis.," Springer-Verlag Berlin Heidelberg ISPEC. LNCS **6047**, 341–358 (2010)
40. S. Aggarwal, S.L. Jatin, B. Garg, Forecasting production values using fuzzy logic interval based partitioning in different intervals. Int. J. Adv. Comput. Sci. Appl. **8**, 292–299 (2017)., **ESCI index**
41. X. Yina, W. Lua, J. Hong, Z.J. Chen, W. Liua, Reversible data hiding in binary images by flipping pattern pair with opposite center pixel. J. Vis. Commun. Image Represent. **70**, 102816 (2020). https://doi.org/10.1016/j.jvcir.2020.102816
42. F. Li, L. Zhang, W. Wei, Reversible data hiding in encrypted binary image with shared pixel prediction and halving compression. J. Image Video Proc. **2020**, 33 (2020). https://doi.org/10.1186/s13640-020-00522-6

Private and Secure Blockchain-Based Mechanism for an Online Voting System

S. Durga, Esther Daniel, S. Seetha, and S. Deepakanmani

Abstract Current electronic voting systems are prone to a range of security threats, including distributed denial of service attacks, polling booth arrangement, vote modification and manipulation, malware attacks, and so on, and they also incur huge organizational resources and time. Existing systems become distrustful as a result. This paper proposes a private and secure blockchain-based mechanism that improves the overall resilience of e-voting. A decentralized and distributed ledger was developed to provide a certain amount of privacy and security for an election by removing any type of tampering. Furthermore, cryptographic principles and a consensus protocol were used to ensure the legitimacy and legality of each transaction, making it safe and thereby strengthening the election process. Every vote cast by a voter through our website is added to the blockchain as a transaction, and the voter is represented by a single unique address allocated to him or her for voting. The proposed mechanism's performance in terms of security, privacy, and scalability reveals that it reduces interference that could taint the integrity of a fair election.

Keywords Blockchain · Online voting · Private and secure · Solidity · Ganache · Truffle · MetaMask

---

S. Durga (✉) · S. Deepakanmani
Sri Krishna College of Engineering and Technology, Coimbatore, India

E. Daniel
Karunya Institute of Technology and Sciences, Coimbatore, India

S. Seetha
CMR Institute of Technology, Bengaluru, India

A. Haldorai et al. (eds.), *4th EAI International Conference on Big Data Innovation for Sustainable Cognitive Computing*, EAI/Springer Innovations in Communication and Computing, https://doi.org/10.1007/978-3-031-07654-1_6

1 Introduction

Voting is an essential fundamental building block towards achieving democracy. Now we have all witnessed the build-up, execution, and culmination of an election in the country, be it over the news, social media, or other forms of media. This has also given us an idea of all the issues and controversies revolving around the result phase where several things such as entrance poll and exit poll stir the boiling pot, creating massive speculation and eventually the election being called rigged by many voters. In India, our voting mechanism consists of an EVM and a VVPAT. Now an electronic voting machine or EVM enables a registered voter to visit a nearby booth and cast his vote by pressing a button for his/her favored candidate and the electronic ballot generates a voter-verified paper audit trail or VVPAT which allows a voter to verify that their vote was cast correctly and also to detect possible election fraud and also serves as a means to audit the EVM-stored election results. Although we have come a long way from when the first EVM election was held, to this day, there are still some prevailing issues such as machine malfunction which could lead to inaccurate results; there is also the issue of the time gap between the voting and the counting of votes which poses a risk of possible tampering, as the ballots are physically stored after votes; "booth capturing" (goons supporting a political party raiding polling stations and stuffing ballot boxes with votes) in some inadequately supervised areas, are some of the issues faced by the Election Commission of India.

1.1 Problem Statement

The country's digital growth, combined with the coronavirus pandemic, has increased the percentage of online users across India. Furthermore, smartphone technology allows anyone with a mobile phone to access the outside world [9]. A secure online voting system is critical in the development of a legitimate voting system. The design of a blockchain-based voting system is essential to overcome problems with traditional voting models and to create qualities of transparency, immutability, and traceability.

1.2 Objectives

In this paper, a distributed ledger based on a smart contract is deployed on a secure blockchain. (i) An election system should be resistant to scenarios of coerced voting. (ii) An election system should not enable traceability of a vote to a voter's identifying credentials which are to provide voter privacy and anonymity. (iii) An election system should provide proof of vote to a voter that the vote was counted correctly. (iv) An election system should not be vulnerable to any form of tampering by a third party. (v) An election system should not be entirely dependent upon a single entity for tallying. (vi) An election system should only allow legitimate, registered members of the electoral rolls to vote. Our solution aims to address these with efficiency and fluidity, to create a more secure but transparent e-voting mechanism. This makes it difficult for the chain to be tampered with since it's secured with cryptography; the perpetrator will have to change the number of blocks that will be caught. Each voter is provided with a private key to initiate a transaction or, in this case, would be in the form of a vote being cast. Now, this private key also acts as a personal digital signature which should not allow a particular voter to vote more than once. At the same time, this also serves as an alias to provide the voter to remain confidential on the blockchain where only his address serves as an identity. Private keys are provided only to eligible individuals on the electoral rolls. Proof of vote can be equated to the roof of work done and transaction confirmation on the blockchain. Transparency is achieved through the concept of a decentralized ledger which enables us to create a public blockchain similar to how transparent the Bitcoin blockchain is [1, 2]. Through this, all transactions(votes) can be tracked, and any tampering can be detected, and at the same time access to functions such as deploying the smart contract and pulling it can be restricted to the admin of the contract [2, 5]. Harnessing decentralization using a distributed ledger called blockchain including systematic tracking with digital signatures (voter private key), transparency of the election process (voter address and vote cast), security (confidentiality and integrity), and the added convenience of voting online installs confidence in the voter and could also reduce the effect of external factors on the elections of our country and pave the way for a more coherent and dependable election [7].

The rest of the paper is organized as follows. Section 2 summarizes the related works. Section 3 explains the private and secure blockchain-based mechanism. Section 4 describes the implementation analysis, and Sect. 5 summarizes the conclusion.

2 Related Work

The current state of blockchain technology and its implementations in the voting scenario are appraised. Blockchain has primarily been used in the financial sector, but due to it being an immutable ledger, it can be used in any process that requires data to be recorded in a timeline and maintain the recorded data's integrity [6]. Multiple research papers highlighting various attempts to successfully implement a voting system that is backed by a blockchain were explored. The various case studies collected and gleaned ranged from developed nations such as Switzerland to smaller nations such as Turkey were to gauge the effectiveness of such systems and critically analyze them to come up with an architecture that would satisfy our requirements. The most successful case seen until now took place in South Korea was in the year 2017; they were able to successfully conduct a blockchain-backed election. Various other degrees of successful implementation have been observed in Iceland, Turkey, and Estonia [7].

The blockchain technology is treated as a service, particularly as an append-only ledger that could be used to record immutable transaction data on it. The implementations of said research ranged from successful elections to just proposals for improvement. A common feature observed was that every voter was authenticated using their government-issued ID, hence allowing only eligible voters to cast a vote [7, 10]. This allowed the government to simply reuse the existing user identification system and not invest in a new database for the eligible voters. Instead, any citizen who has attained the minimum voting age is simply marked as an eligible voter and is allowed to vote in the upcoming elections. Yet the process involved in the various cases was a lot different. In most cases, the vote would be a variable or attribute but in one case, the actual vote cast was proposed to be the cryptocurrency value involved in a transaction. This presented a whole new level of transparency and traceability as each vote call would also be recorded on the blockchain [7, 8, 10]. This idea, although requiring more transactions on the blockchain, would ensure that every vote is recorded on the network itself, and any vote cast can be verified using the transaction ID obtained from a successful transaction.

It is observed from the literature that these existing systems stand divided is the actual architecture of the blockchain. In the case of Iceland, the team suggested a single big blockchain network whose nodes were the actual district ballots throughout the country. Any vote cast will have to reach a consensus in this giant network which might pose a delay. To counter this, a different architecture is proposed that involves a hierarchical arrangement of various chains, wherein each chain represents a ballot station and, in some cases, an entire district [11]. But, this would add complexity to the network; a consensus would have to be reached multiple times when one transaction is recorded. Hence, it may cause congestion in the network.

The authentication of the voters can also be achieved through various methods, the most common being the verification of the individual using the help of an ID verification API. This API could be provided by a security firm mandated by the government or by the government directly [7]. In countries where a unique identifier system exists, such as the social security number in the United States of America or the AADHAAR in the Indian subcontinent, these system's APIs could be used. In other countries, a new verification API would have to be developed.

Alternatively, a database containing the credentials of all eligible voters would be managed by the election commission involved and be updated annually and verified before any election run. Some models retained the verification method from the traditional elections, i.e., in-person verification, which involved the voters producing their voter ID or SSN ID(s) to verify their eligibility [8, 9]. But this would impose a higher cost for the election, due to the cost of the technological needs and the manpower required to carry out the verification manually. Ideally, it is not suitable for our country and has to be avoided as India already hosts one of the most expensive elections and adding to this enormous sum might not be favorable.

It is observed that although various teams have been researching a possibility of such an efficient voting system, it remains in rudimentary stages. And the reason for this state is different in each scenario, from huge implementation costs to cynical views of blockchain technology; many countries, although capable, hesitate to transition to such a system. In this paper, an architecture that can be implemented easily and be used successfully to conduct any type of election is implemented. An architecture that will minimize the risks involved and will also guarantee a tamper-free election process is developed. The Internet threats such as botnets and other adversarial attacks on the e-voting system have to be efficiently identified. The early detection of the botnet in the e-voting system is implemented using the evolutionary algorithms and swarm techniques. As future work, the proposed E-voting protocol will incorporate efficient decision-making strategy by using evolutionary algorithms to accurately identify the botnets and a variety of other attacks [14].

3 Private and Secure Blockchain-Based Mechanism

3.1 Block Diagram of Private and Secure Blockchain-Based Mechanism

The block diagram of the private and secure Blockchain-based mechanism is shown in Fig. 1 which includes seven different components described as follows:

1. **Voter Authentication** – A method for eligible voters to gain access to the voting interface. A voter is to input his credentials and authenticate themselves so that they can proceed to vote for their desired candidate. A voter is deemed eligible using their personally identifiable information that is fetched from a service like AADHAAR.
2. **Voter Database** – A collection of personal and demographic information of all eligible voters. When a voter tries to authenticate themselves, their information is checked and then fetched from this database.
3. **Metamask Wallet** – The Ethereum wallet from which the voter casts the vote.
4. **Voting Interface** – The web portal is used to cast votes and view results.
5. **Election Contract** – The deployed instance of a solidity contract that is called whenever a transaction on the blockchain is to be initiated. The contract controls the main aspects of our election such as candidate registration, voting, results aggregation, etc.
6. **Admin Panel** – The admin panel helps the election deployer manage and monitor the entire election. An administrator can use the portal to register candidates, add users, add wallet addresses, and even view results.
7. **Results Page** – A webpage dedicated to showcasing the statistics of the current election. Involves multiple charts to provide a clear view of the state of the election.

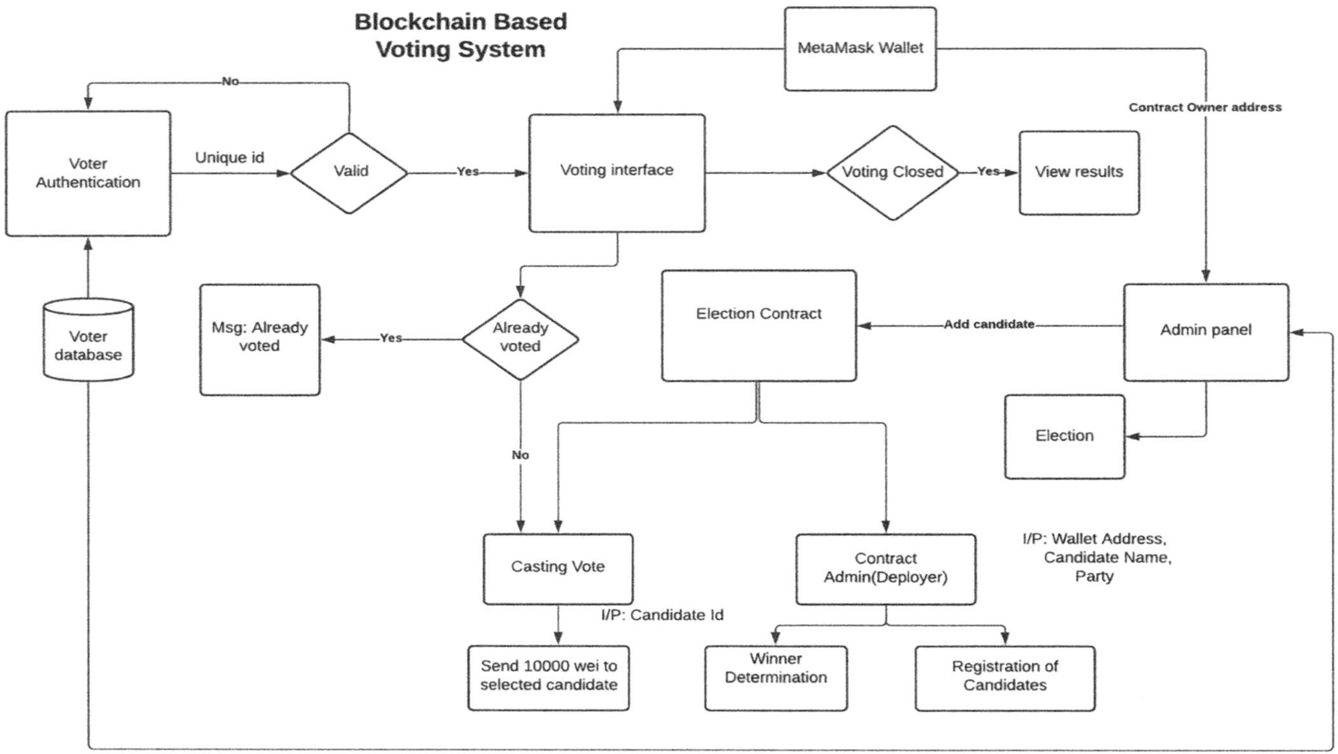

Fig. 1 Block diagram of private and secure blockchain-based mechanism Voting Process Diagram

These components function together to enable the voter to vote successfully, and the algorithm below gives a general idea of how the system works.

3.2 *Voting Process Diagram*

When the Election Commission deploys the contract, it transmits one from the owner's wallet, which is added to the contract balance. When a user votes, 10,000 Wei is sent to the candidate address balance of the candidate for whom the user voted. The election will be won by the candidate with the highest balance. Figure 2 depicts the transactions that occur during the election process.

Figure 3 shows the sequence diagram that describes the sequence of steps the system goes through when a voter casts a vote.

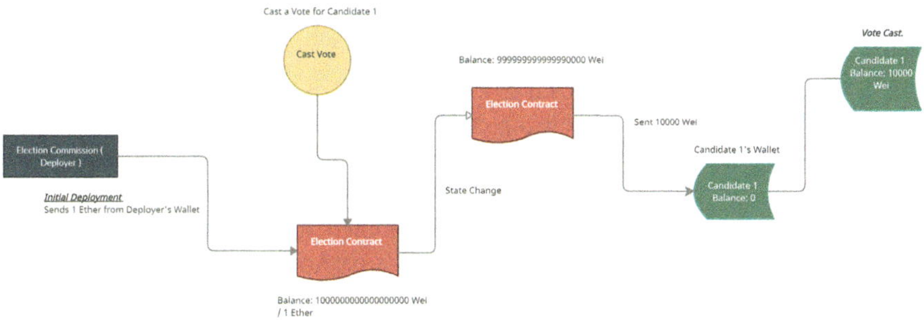

Fig. 2 State diagram with contract and the wallets associated in the election

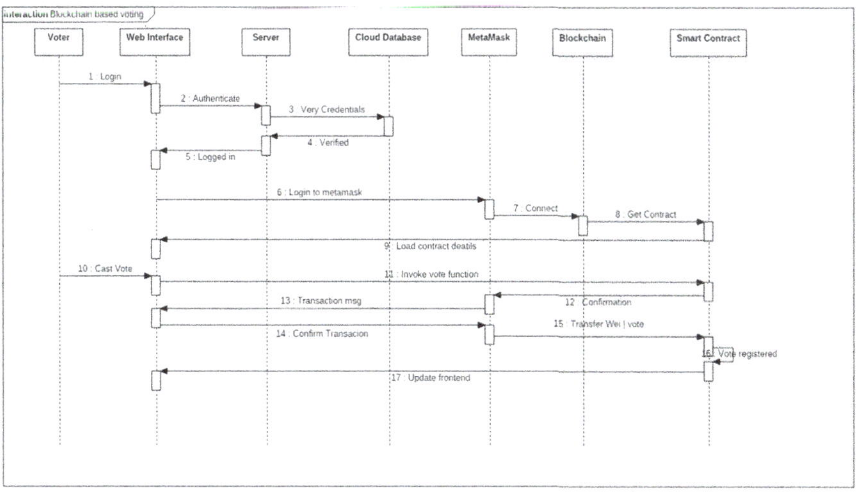

Fig. 3 Sequence diagram

3.3 *Major Entities*

The decentralization of a voting system involves the action of casting a vote to be recorded on an immutable ledger, i.e., blockchain. The votes are cast through an entity called the smart contract which in layman's terms is a definite piece of code that is employed on the blockchain, and this can be invoked through its deployed address to perform a transaction on the blockchain. Through this transaction, a voter's vote is cast. Later, the administrator or deployer of the election smart contract can retrieve the results.

The major entities involved in the proposed mechanism are:

(a) The blockchain network.
(b) Smart contract.
(c) Ethereum wallets.

3.3.1 The Blockchain Network

The Blockchain, previously known as 'block chain' is a growing list of records, called blocks, that are connected using cryptography. Each block contains an SHA-256 hash of the preceding block, a timestamp, and some form of transaction data [1]. A blockchain is resistant to any form of tampering with its data. This is because once recorded, the data in a block cannot be altered without the alteration of all the descendant blocks, since the previous block's hash value has to be changed for every block that follows the block that has been modified. In addition to this, the entire copy of the current blockchain network is distributed among its nodes. This inoculates the blockchain against the addition of malicious blocks in the network [1]. Hence, in this study, we chose to use such an immutable distributed ledger to be the base of the voting system, as this can potentially eradicate the issues of voter fraud. In addition, it can also provide transparency and traceability in the voting system. We have chosen to use the Ethereum blockchain particularly due to its ability to house smart contracts and also its immense support for solidity [2].

3.3.2 Smart Contract

A smart contract is a self-executing contract with the terms of an agreement between the parties being directly mandated into lines of code. The code and the agreement stipulated by this contract are present across a distributed, decentralized blockchain network. The code controls all aspects of the contract including deployment execution and error reporting. The transactions which occur are also trackable and irreversible [2, 3]. Although normally smart contracts would be used in situations where a sale is being made and some conditions are to be satisfied to proceed with the transaction, they can be modified to suit various needs. And in this voting scenario, they will be used to deploy the election, register the candidates, cast votes, tally the votes, and finally publish the results. Our entire election process is boiled down to a few lines of code in a smart contract. These smart contracts are written in a language known as solidity [13]. (Solidity is a contract-oriented, high-level language for implementing smart contracts.)

3.3.3 Ethereum Wallets

An Ethereum wallet is a digital wallet that allows the user to have and perform transactions with their ether. Every wallet has a mnemonic and its associated private which is generated from the mnemonic. This private key (also called master private key) can be used to generate multiple public keys which refer to the specific wallet. And this public key is shared with others so that the user can receive ether from other wallet holders [2, 4].

In our study, these wallets primarily denote our voters and candidates. Voters will be provided with a wallet with a certain amount of ether. This ether can be sent to the candidates' wallets. This transaction from the voter's wallet to the chosen candidate's wallet is considered a vote.

When results are to be compiled, the contract simply fetches the wallet balances of the candidates' wallets and compares them with its peers to find the candidate with the highest wallet balance. Since the candidate with the highest ether balance should have received the highest number of votes, they will be declared as the winner. Hence, using these technologies available to us, we will be implementing our proposed system.

Algorithm

1. The election commission deploys the election contract.
2. The Election Commission registers the candidates for each constituency and declares the voting window.
3. The Election Commission emails every eligible voter with their credential pair.
4. The User will log in to the website and vote and has to be authenticated first using the Voter Database.
5. Voter Authentication module pulls the voter details from the database to compare with the credentials entered by the user.

```
Input: User Unique ID and Password
if (Credentials supplied exists in the voter database) {

  Allow Login;
  Fetch user details && Redirect to voting interface;

}
else {

  Deny Login;
  Show error && Request user to try again;

}
```

6. The Voting Interface module is available as long as the deadline specified in the smart contract does not expire.

   if (current_time! = deadline_time) && (voter has not voted yet) {
   Fetch candidates based on candidate's constituency;

     Wait for candidate selection;

   when (candidate is selected) {

     Cast_Vote(Candidate_ID){

   Obtain candidate's wallet address from election contract;
   Start a transfer 10,000 Wei to candidate's wallet address;
   Obtain confirmation from voter in metamask dialog;
   Confirm the transaction and obtain transaction receipt;
   Voters [voters_length + 1] = current_voter_wallet_address;

     Redirect voter to results page;

       }

   }
   else if (current_time = > deadline_time) || (voter has already voted){

     Block voting capability;
     Redirect to results page;

   }
   else {.

     Show error and redirect to the user login page;

   }

7. if AlreadyVoted(true) then, display a message to inform the voter they cannot vote again and hide the Casting Vote module.

   else, users can select a candidate from their constituency, to vote for.

8. MetaMask login will be required for the Casting Vote module. A private key is supplied to the voter by the Admin Panel (Contract Owner) module via email. This private key allows you to connect to MetaMask Wallet which provides you with the ability to cast a vote.

   Input: Metamask Wallet Private Key
   Process: Wallet key verification
   Output: Redirect voter to a user authentication page

9. The Casting Vote sub-module which is a part of the Election Contract module allows the voter to cast a vote provided certain conditions are met.
10. if hasNotVoted(true) && validCandidateID(true) && validAddress(true) && votingDeadlineExpired(false) then, utilizing a function which takes the Candi-

date ID as a parameter it updates the balance in the candidate's wallet, sending 10,000 wei to the selected candidate's account and the voter status is recorded to disallow more than one vote from the same person.

else, an error message about the error in question will be displayed to the Voter.

11. The Admin Panel is responsible for authenticating voters as seen in step 3, populating the Voter Database, and is also connected to the Election Contract module.
12. The Contract Admin sub-module is also responsible for deploying the contract (truffle migrate – reset) and for adding the candidates to the Election Contract module. To register a candidate, the admin needs to provide the Candidate-WalletAddress, CandidateName, CandidateParty, and CandidateConstituency as necessary parameters to the function whose access specifier is strictly limited to the owner.
13. To display winners the Contract Admin has a sub-module, Winner Determination, which is the calling of the GetWinners() function of the contract which will calculate the balances for each candidate address (accumulated by the transfer of 10,000 Wei for each vote cast to the candidate), and the candidate with the highest balance is declared as a winner.

```
for each (constituency in election) {
    Winner = candidate_id with maximum votes in constituency
    Add Winner to winners array
}
```

14. Once the deadline for the contract has passed, indicating voting has ended, the Admin Panel will withdraw the deployed Election Contract, and no further transactions will be allowed to take place.

4 System Implementation

This section describes the implementation of the proposed mechanism. The Ethereum blockchain is used due to its ability to house smart contracts and also for its immense support of solidity domain-specific language [15–17]. The use of frameworks such as Ganache and Truffle [12] has been explored, in the development of smart contracts and their deployment. Linking the smart contract to a web interface has been done to cast the votes in an interactive and user-friendly manner. The proposed election mechanism involves the deployment of a smart contract on the blockchain. For this, we needed to come up with a design of a smart contract that would be able to provide us with a candidate model and also perform the functions mentioned above. Ganache, Truffle, Solidity, and MetaMask are the four software used in the implementation.

The details of the software involved in building the application are as follows:

1. Ganache – This provides us with a custom RPC network on localhost port 7545 for managing and monitoring our blockchain. It also provides us with some dummy addresses and private keys with an initial balance of a certain amount of ether in each address which is obviously, not of real value but merely for testing. In addition to this, it also shows information regarding transactions and the blocks involved in our blockchain.
2. Truffle – Sophisticated framework for the development and testing of our smart contract. It helps in lifecycle management, automatic testing, as well as deploying the smart contract to our local blockchain. It also provides an interactive and powerful console that saves time when interaction with the contract is needed.
3. Solidity – Object-oriented programming language for building smart contracts. It can be used to run smart contracts on various blockchains including the one we are using, Ethereum.
4. MetaMask – This is a widely used Ethereum wallet manager which is essential for carrying out transactions via your browser. It can be added as an extension to the browser. The private key is used to create an account on MetaMask, and only then can a voter cast their vote through our website to their desired candidate. They will receive a transaction summary and confirmation if the vote has been cast successfully.

4.1 Security Analysis

A frequent network attack that can be used against the distributed web application is the DDoS (distributed denial-of-service) attack [18]. The proposed mechanism has aimed to mitigate the DDoS attack. The attacker must DDoS every boot node in the private network. If this occurred, the person or institution would be quickly identified. Each node has a Byzantine fault-tolerance technique, which aids in the location of failed nodes in the system. By enabling confidentiality through "public key infrastructure" that guards against malicious efforts to alter data and preserving the size of a ledger, the proposed mechanism can achieve concurrent security and privacy. A voter must enter his or her credentials such as unique PIN, voter ID, etc. and confirm themselves before voting for their preferred candidate. A voter's eligibility is determined using personal identifying information obtained from a service such as AADHAAR. The voter authentication module retrieves voter information from the database and compares it to the user's credentials. If an individual knew the PIN for each corresponding electronic ID he had, he could vote for many people without monitoring. A biometric scan could be implemented shortly to further address this problem.

5 Conclusion

In this paper, a blockchain-based mechanism for a secure electronic voting system is implemented. The smart contract has been used to make elections more safe and cost-effective while maintaining voter privacy. The systems architecture, the design, and the security analysis are elaborated. The main purpose of developing a web-based voting application is to decentralize the process of voting and harness the best features of a decentralized blockchain. This reduces the costs of the election process considerably and also reduces tampering of results. It increases confidence in democracy as the transparency of the blockchain while maintaining voter privacy and promotes a well-informed election that is free from any physical impediments and human interferences. A transaction-based voting to record every vote as a block on our Ethereum private blockchain is implemented. Blockchain is fundamentally strengthened by cryptography, decentralization, and consensus, so any changes made to the blockchain will indeed be detected and recorded, and any unauthorized intrusion will be prevented.

References

1. L. Schilling, H. Uhlig, Some simple bitcoin economics. J. Monet. Econ. **106**, 16–26 (2019)
2. A.S. Andreou, P. Christodoulou, K. Christodoulou, A decentralized application for logistics: Using blockchain in real-world applications. Cyprus Rev. **30**, 181–193 (2018)
3. X. Li, P. Jiang, T. Chen, X. Luo, Q. Wen, A survey on the security of blockchain systems. Futur. Gener. Comput. Syst. **107**, 841–853 (2020)
4. M. Alharby, A. Aldweesh, A. van Moorsel, *Blockchain-Based Smart Contracts: A Systematic Mapping Study of Academic Research* (School of Computing, Newcastle University, Newcastle upon Tyne, 2018)
5. B. Wang, J. Sun, Y. He, D. Pang, N. Lu, Large-scale election based on blockchain. Procedia Comput. Sci. **129**, 234–237 (2018)
6. P.M. Dhulavvagol, V.H. Bhajantri, S.G. Totad, Blockchain ethereum clients performance analysis considering E-voting application. Procedia Comput. Sci. **167**, 2506–2515 (2020)
7. A. Dhillon, G. Kotsialou, P. McBurney, L. Riley, Introduction to voting and the blockchain: some open questions for economists (No. 416). Competitive Advantage in the Global Economy (CAGE). (2019)
8. R. Bulut, A. Kantarcı, S. Keskin, Ş. Bahtiyar, Blockchain-based electronic voting system for elections in turkey, in *2019 4th International Conference on Computer Science and Engineering (UBMK)*, (IEEE, 2019), pp. 183–188
9. S. Durga, S. Surya, E.S. Daniel, MobiCam: Towards a new paradigm for leveraging smartphone cameras and IaaS cloud for smart city video surveillance, in *2018 2nd International Conference on Trends in Electronics and Informatics (ICOEI)*, (IEEE, 2018), pp. 1035–1038
10. F. Casino, T.K. Dasaklis, C. Patsakis, A systematic literature review of blockchain-based applications: Current status, classification and open issues. Telematics Inform. **36**, 55–81 (2019)
11. K.M. Khan, J. Arshad, M.M. Khan, Secure digital voting system based on blockchain technology. Int. J. Electron. Gov. Res. **14**(1), 53–62 (2018)
12. Truffle Docs. https://www.trufflesuite.com/docs/truffle/overview
13. Solidity Docs. https://docs.soliditylang.org/en/v0.8.3/

14. M. Asadi, M.A.J. Jamali, S. Parsa, V. Majidnezhad, Detecting botnets by using particle swarm optimization algorithms based on voting systems. Futur. Gener. Comput. Syst. **107**, 95–111 (2020)
15. U. Jafar, M.J.A. Aziz, Z. Shukur, Blockchain for electronic voting system – Review and open research challenges. Sensors **21**(17), 5874 (2021)
16. M. Ibrahim, K. Ravindran, H. Lee, O. Farooqui, Q.H. Mahmoud, Election block: An electronic voting system using Blockchain and fingerprint authentication, in *2021 IEEE 18th International Conference on Software Architecture Companion (ICSA-C)*, (IEEE, 2021), pp. 123–129
17. S.P. Gupta, A.M. Tripathi, E-voting using Blockchain. J. Phys. Conf. Ser. **1911**(1), 012001). IOP Publishing (2021)
18. N. Indrason, W. Ksshongbuh, G. Saha, Blockchain-based boothless e-voting system, in *International Conference on Innovative Computing and Communications*, (Springer, Singapore, 2021), pp. 779–788

A Novel Real-Time 3D Object Detection Network in Autonomous Driving Using Reformed RS-ResNet network

Sai Prabanjan Kumar , C. Mala , and V. Punitha

Abstract Extending the RS-ResNets architecture for 3D bounding box detection from 2D images to 3D point clouds that have varying degrees of visibility, a regression module with a Euler Region Proposal Network that performs complex angle regression is added to predict the direction of bounding boxes around detected objects. Geometrical overlap of predicted labels and ground truth is identified for every set of Lidar point cloud and the corresponding image data and compared with the state-of-the-art architectures. With the ability to scale to different architectures of GPU/TPU, the RS-ResNet architecture achieves faster frame rates of reaching threshold value of 70 ms/frame and APH/L2 = 66, on a NVIDIA Tesla V100 GPU and capability to scale to new and more powerful GPU/TPU architectures. The predictions are made to two levels of label hierarchy and three levels of Lidar range images with the associated camera images that are considered as single set. The achieved frame rates and detected AP are compared to the latest state-of-the-art architectures such as Complex-YOLOV4 and other leaderboard in the Waymo 2021 3D object detection challenge.

Keywords RS-ResNet · YOLOV4 · Euler Region Proposal Network · 3D bounding box detection

1 Introduction

With autonomous vehicles, the requirement of a human driver for safely operating a vehicle is no longer required, as a combination of sensors such as lidar, radar, and camera along with a software stack can do the same task. Combining different

S. P. Kumar (✉) · C. Mala
National Institute of Technology, Tiruchirapalli, India
e-mail: mala@nitt.edu

V. Punitha
Saranathan College of Engineering, Tiruchirapalli, India

A. Haldorai et al. (eds.), *4th EAI International Conference on Big Data Innovation for Sustainable Cognitive Computing*, EAI/Springer Innovations in Communication and Computing, https://doi.org/10.1007/978-3-031-07654-1_7

sensors so that the weakness of one sensor can be complemented by another sensor, and this process is called as sensor fusion.

Detecting objects in camera images can be done using deep neural networks, but for enabling autonomous driving high levels of precision are required to ensure objects are detected independent of sensor failures due to weather or lighting conditions. Sensor fusion enables lidar data and image data to include lidar data to improve the tracking results through sensor fusion. Additional sensors such as radars and lidars have different physical measurement principle and improve the reliability of the environment perception system.

Lidar sensor generates point cloud using the sensor that uses light to measure distance, the sensor rotates on its axis and generates 1.5 million data points per second, but based on this 3-D point clouds alone, autonomous vehicles cannot safely navigate. In order to make decisions and perform path planning or issue braking maneuvers, an autonomous vehicle needs to identify relevant objects in its surroundings. Such objects are other vehicles of various types (e.g., cars, trucks, trailers), cyclists, pedestrians, lane boundaries, road signs, and several others. In computer vision, deep-learning approaches are often used to detect and classify relevant objects in a scene. Objects will be classified as pedestrians, vehicles, and cyclists, and road signs have been detected and correctly classified by the YOLO detection framework. Until recently, 2D image processing was the major domain, in which detection and classification based on deep learning were performed. Deep-learning methods applied on 3D point clouds are having both academic importance and business importance with applications in automobile industry and satellite data processing. The state-of-the-art methods for processing the problems relating to point cloud processing has increased because of the same. But adopting the deep-learning methods from image or computer vision to point cloud data is having its own problems.

1.1 Adopting Image Data to Lidar Data-Related Problems

1. Point clouds are of a sparse and unstructured nature as the location of points in a scene, and therefore, both their density and number vary with the presence of objects in a scene. An image taken by a camera on the other hand always has the same number of pixels that can be fed to a neural network.
2. As autonomous vehicles need to react very quickly, object detection must be performed in real time. This means that a detection network must provide results within the time interval in between two scans, trying to understand processing pipeline for object detection and classification based on point clouds [6, 10].

1.1.1 Major Parts in an Object Detection Pipeline

Any processing pipeline for object detection and classification consists of three major parts:

(1) Representation of data
(2) Extracting features from data
(3) Detecting based on model performance

In the first part, the point cloud provided by the lidar sensor is organized into a structure that is suited for an efficient processing of the data in the subsequent pipeline stages. In the existing research, two major approaches for data representation exist: either the point cloud is transformed into a structure such as "voxels," "pillars," or "frustums" or the point cloud is left as is and processed directly. Both approaches have their respective advantages and disadvantages. In the second part, features are extracted in low dimensional and high dimensional characteristics, similar to feature extraction in images. The feature extraction is performed on transformed point cloud instead of 3D point cloud in order to reduce the computational load, where the 3D point cloud is represented as a bird's eye view using grid representation. Once the features are extracted, these extracted features are forwarded to the object detection network. Object detection network predicts bounding boxes as regression output using pixel-wise labeling, along with predicting the object class. And by using the Euler-Region Proposal Network the orientation and rate of change of direction will be predicted.There are a lot of state-of-the-art three-step object detection pipelines present that incorporate or adopt the image object detection data to point cloud data.

Steps in Clustering the Point Cloud Dataset The steps in the clustering pipeline can be broadly explained as:

Step 1: Data representation for easy processing of the point cloud data: Lidar point clouds are an unstructured assortment of data points that are distributed unevenly over the measurement range. With the prevalence of convolution neural networks (CNN) in object detection, point cloud representations are required to have a structure that suits the need of the CNN, so that convolution operations can be efficiently applied. The available methods can be broadly explained as: point-based data representation. Point-based methods take the raw and unfiltered input point cloud and transform it into a sparse representation, which essentially corresponds to a clustering operation, where points are assigned to the same cluster based on some criterion (e.g., spatial distance).
Step 2: Spatial distance is the important factor in clustering the point cloud approach, where extracting a feature vector for each point in a point cloud is done by considering the neighboring clusters [1].

1.2 Projection of Lidar Range Data

The traditional approach for this method is to take every single point's low-dimensional characteristics and then aggregate them to larger and more complex high-dimensional features. One of the most prominent representatives of this class of approaches is PointNet by Qi et al., which has in turn inspired many other significant contributions such as PointNet++ or LaserNet. One of the major advantages of point-based methods is that they leave the structure of the point cloud intact so that no information is lost, e.g., due to clustering. However, one of the downsides of point-based approaches is their relatively high need for memory resources as a large number of points have to be transported through the processing pipeline. Voxel-based data representation: A voxel is defined as a volume element in a three-dimensional grid in space. A voxel-based approach assigns each point from the input point cloud to a specific volume element. Depending on the coarseness of the voxel grid, multiple points may land within the same volume element. Then, in the next step, local features are extracted from the group of points within each voxel. One of the most significant advantages of voxel-based methods is that they save memory resources as they reduce the number of elements that have to be held in memory simultaneously. Therefore, the feature extraction network will be computationally more efficient because features are extracted for a group of voxels instead of extracting them for each point individually. A well-known representative of this class of algorithms is VoxelNet. The following figure shows a point cloud whose individual points are clustered based on their spatial proximity and assigned to voxels. After the operation is complete, the amount of data representing the object has significantly decreased. Pillar-based data representation: An approach very similar to voxel-based representation is the pillar-based approach. Here, the point cloud is clustered not into cubic volume elements but instead into vertical columns rising up from the ground up. As with the voxel-based approach, segmenting the point cloud into discrete volume elements saves memory resources—even more so with pillars as there are usually significantly fewer pillars than voxels. A well-known detection algorithm from this class is PointPillars. Frustum-based data representation: When combined with another sensor such as a camera, lidar point clouds can be clustered based on pre-detected 2D objects, such as vehicles or pedestrians. If the 2D region around the projection of an object on the image plane is known, a frustum can be projected into 3D space using both the internal and external calibration of the camera. One method belonging to this class is, e.g., Frustum PointNets. The following figure illustrates the principle. One obvious disadvantage of this method when compared to the previous ones is that it requires a second sensor such as a camera. However, as these are already used for object detection in autonomous driving and guaranteed to be on-board a vehicle, this is not a significant downside. In projection-based data representation, while both voxel- and pillar-based algorithms cluster the point cloud based on a spatial proximity measure, projection-based approaches reduce the dimensionality of the 3D point cloud along a specified dimension. In the literature, three major approaches can be identified, which are front view (RV), range view (RV), and bird's eye view (BEV). In the FV approaches, the point cloud is compacted along the forward-facing axis, while with BEV images, points are projected onto the ground plane. The following figure illustrates both methods. RV methods are very similar to the FV approach with the exception that the point cloud is not projected onto a plane but onto a panoramic view instead. This concept is implemented in datasets such as Waymo dataset, in which lidar data is stored as range images. In the literature, BEV is the projection scheme most widely used [2, 8].

1.2.1 Reasons for BEV Popularity

The reasons for this are three-fold:

(1) The objects of interest are located on the same plane as the sensor-equipped vehicle with only little variance.
(2) The BEV projection preserves the physical size and the proximity relations between objects, separating them more clearly than with both the FV and RV projection.
(3) Feature extraction: After the point cloud has been transformed into a suitable representation (such as a BEV projection), the next step is to identify suitable features. Currently, feature extraction is one of the most active research areas, and significant progress has been made there in the last years, especially in improving the efficiency of the object detector models.

The proposed method of the paper takes the progress made in the field into consideration, and after comparing with recent state-of-the-art techniques, the method proposed utilizes the latest improvements in the training of the models and achieved comparable results with the existing state of the art along with greater hardware interoperability. Also the existing state-of-the-art models work by converting image data using stereo camera images for 3D object detection that is susceptible to adversarial attacks, but the proposed method learns 3D object detection using Lidar data as a sensor modality, which makes the inference risk of attacks by significant value.

The proposed model also detects data falsification by the approach of false positive data calculations that have been explained in the proposed methodology. False positives and false negatives are utilized for identifying false data being taken into intersection over union calculations.

2 Related Work

The type of features that are most commonly used are: (1) local, (2) global, and (3) contextual features: 1. Local features, which are often referred to as low-level features, are usually obtained in a very early processing stage and contain precise information, e.g., about the localization of individual elements of the data representation structure. 2. Global features, which are also called high-level features, often encode the geometric structure of an element within the data representation structure in relation to its neighbors. 3. Contextual features are extracted during the last stage of the processing pipeline. These features aim at being accurately located and having rich semantic information such as object class, bounding box shape and size, and the orientation of the object. In the following, we will look at a number of feature extractor classes found in the current literature: Point-wise feature extractors. The term "point-wise" refers to the fact that the entire point cloud is used as input [5]. This approach is obviously suited for the point-based data representation from the first step. Point-wise feature extractors analyze and label each point individually, such as in PointNet and PointNet++, which currently are among the most well-known feature extractors. To illustrate the principle, let us briefly look at the PointNet architecture, which is illustrated in the following figure: PointNet uses the entire point cloud as input. It extracts global structures from spatial features of each point within a subset of points in Euclidean space. To achieve this, PointNet implements a non-hierarchical neural network that consists of the three main blocks, which are a max-pooling layer, a structure for combining local and global information and two networks that align the input points with the extracted point features. Finally, a three-layer fully connected network is used to map the global feature vector to generate both object classification and object location. One of the downsides of PointNet is its inability to capture local structure information between neighboring points since features are learned individually for each point and the relation between points is ignored. Even though point-wise feature extractors show very promising results, they are not yet suitable for use in autonomous driving due to high-memory requirements and computational complexity. Segment-wise feature extractors: Due to the high computational complexity of point-based features, alternative approaches are needed so that object detection in lidar point clouds can be used in a real-time environment. The term "segment-wise" refers to the way how the point cloud is divided into spatial clusters (e.g., voxels, pillars, or frustums). [5] Once this has been done, a classification model is applied to each point of a segment to extract suitable volumetric features. One of the most-cited representatives of this class of feature extractors is VoxelNet. In a nutshell, the idea of VoxelNet is to encode each voxel via an architecture called "Voxel Feature Extractor (VFE)" and then combine local voxel features using 3D convolutional layers and then transform the point cloud into a high-dimensional volumetric representation. Finally, a region proposal network processes the volumetric representation and outputs the actual detection results. Convolutional neural networks (CNN): For years, CNN have been used successfully to detect and classify objects in camera images. One of the first landmark papers, which showed that CNN-based image classification outperformed the back-then state-of-the-art methods such as support vector machines (SVM), was published in 2012 by Krizhevsky et al. In recent years, many of the approaches for image-based object detection have been successfully transferred to point cloud processing. In CNN-based object detection methods, the processing pipeline relies heavily on the use of "backbone networks," which serve as basic elements to extract features. This approach allows for the adaptive and automatic identification of features without the need to invest manual (and thus often heuristic) engineering efforts as with many classic approaches. In most cases, the backbone networks used for image-based object detection can be directly used for point clouds as well. In order to balance between detection accuracy and efficiency, the type of backbones can be chosen between deeper and densely connected networks or lightweight variants with few connections. Even though the results achieved with CNNs such as AlexNet were stunning at the time, these networks had the problem that with the network depth increasing, the accuracy of detection became saturated and degraded rapidly, due to a problem referred to as "vanishing gradients." This problem generally occurred in all architectures with a high number of layers, i.e., in all "deep" networks. To overcome this problem, the ResNet architecture was proposed by He et al. in this paper, which used "skip connections" (i.e., shortcuts) to directly pass values of one layer to the next layer without using a non-linear transformation. By using such shortcuts, gradients can be directly propagated, leading to significant reductions in training difficulty. This means that network depth can be increased without compromising the model's training capabilities [5]. Detection and Prediction Refinement. Once features have been extracted from the input data, a detection network is needed to generate contextual features (e.g., object class, bounding box) and finally output the model predictions. Depending on the architecture, the detection process can either perform a single pass or a dual pass. Based on the detector network architecture, the available types can be broadly organized into two classes, which are dual-stage encoders such as R-CNN, Faster R-CNN, and PointRCNN or single-stage encoders such as YOLO or SSD. In general, single-stage encoders are faster than dual-stage encoders, which make them more suited for real-time applications such as autonomous driving. A problem faced by CNN-based object detection is that we do not know how many instances of a certain object type are located within the input data. It could be that only a single vehicle is visible in the point cloud, or it could also be 10 vehicles. A naive approach to solve this problem would be to apply a CNN to multiple regions and check for the presence of objects within each region individually. However, as objects will have different locations and shapes, one would have to select a very large number of regions, which quickly become computationally infeasible. To solve this problem, Ross Girshick et al. proposed a method (R-CNN) where a selective search is used to extract 2000 regions, which he called region proposals. This meant a significant decrease in the number of regions that needed to be classified. Note that in the original publication, the input data were camera images and not point clouds. The candidate regions are then fed into a CNN to produce a high-dimensional feature vector, from which the presence of objects within the candidate regions is inferred using a support vector machine (SVM). In order to refine the predictions and increase the accuracy of the model output, dual-stage encoders feed the results from the first stage to an additional detection network that refines the predictions by combining different feature types to produce refinement results. Single-stage object detectors on the other hand perform region proposal, classification, and bounding box regression all in one step, which make them significantly faster and thus more suitable for real-time applications. But considering the accuracy values as a parameter, two-stage detectors tend to perform better over the single-stage detectors. One of the most famous single-stage detectors is YOLO (You Only Look Once). YOLO model contains a convolution network for extracting low-level features and high-level features, which are the outcome of network predictions. This low-level and high-level predictions are used for decoding the predictions and generating bounding boxes. Generating bounding boxes is a complicated task that involves YOLO algorithm using pre-defined anchor boxes with a certain aspect ratio. This aspect ratio is a variable that depends on training dataset and the dimensions of the objects present in that dataset [3, 9]. Using anchor boxes removes the need for sliding window operation of the detector that is resource-intensive. Whereas the anchor boxes can be placed on the image and the network predicts the object class probability, bounding box dimensions using the anchor box and all the object predictions and its attribute predictions can be made in a single pipeline pass instead of a sliding window operation, which is used in many state-of-the-art approaches. The approach of anchor boxes is unique, as it enables the processing of entire input data at once and making the prediction of objects and their sliding windows in a single pass. Each anchor box is utilized for predicting the object class and object dimensions, and also the network gets the capability of identifying multiple objects, different scaled objects, and partially exposed objects [4, 7]. As stated before, most of the CNN-based approaches originally come from the computer vision domain and have been developed with image-based detection in mind. Hence, in order to apply these methods to lidar point clouds, a conversion of 3D points into a 2D domain has to be performed.

3 Proposed Methodology

Real-time 3D object detection on point clouds: M. Simon et al. extend the famous YOLO network for bounding box detection in 2D images to 3D point clouds. The main pipeline of Complex YOLO consists of three steps: 1. Transforming the point cloud into a bird's eye view (BEV). First, the 3D point cloud is converted into a bird's eye view (BEV), which is achieved by compacting the point cloud along the upward-facing axis. The BEV is divided into a grid consisting of equally sized cells, which enables us to treat it as an image, where each pixel corresponds to a region on the road surface. As can be seen from the following figure, several individual points often fall into the same grid element, especially on surfaces that are orthogonal to the road surface. The following figure illustrates the concept: As can be seen, the density of points varies strongly between cells, depending on the presence of objects in the scene. While on the road surface, the number of points is comparatively low due to the angular resolution in vertical direction (64 laser beams), the number of points on the front, back, or side of a vehicle is much higher as neighboring vertical LEDs are reflected from the same distance. This means that we can derive three pieces of information for each BEV cell, which are the intensity of the points, their height, and their density. Hence, the resulting BEV map will have three channels, which from the perspective of the detection network makes it a color image. The process of generating the BEV map is as follows: 1. First, the area to encompass needs to be decided. For the object detection in this course, the longitudinal range is set to 0...50 m, and the lateral range to −25...+25 m. The rationale for choosing this particular set of parameters is based partially on the original paper as well as on design choices in the existing implementations of Complex YOLO. 2. Then, the area is divided into a grid either by specifying the resolution of the resulting BEV image or by defining the size of a single grid cell. In the implementation, the size of the BEV image is set to 608 × 608 pixels, which results in a proper spatial resolution. On the top left, you can see the BEV map with all three channels superimposed. On the top right, you can observe the height coded in green. It can clearly be seen that the roofs of the vehicles have a higher intensity than the road surface. 2. Complex YOLO on BEV map. Understanding the network architecture: Instead of a simplified YOLOv2 architecture, an extended YOLOV4 architecture is used. Extensions to the original YOLO network are a complex angle regression and an Euler Region Proposal Network (E-RPN), which serve to obtain the direction of bounding boxes around detected objects. The YOLO network has been configured to divide the image into a 16×32 grid and predicts 75 features. The model has a total of 18 convolutional layers and 5 pooling layers. Also, there are 3 intermediate layers, which are used for feature reorganization. More details on the network layout can be obtained from Table 1. Let us discuss how the features per grid cell are obtained: 1. The YOLO network predicts a fixed set of boxes per cell, in this case 5. For each box, 6 individual parameters are obtained, which are its two-dimensional position in the BEV map, its width and length, and two components for the orientation angle. 2. In addition to the box parameters, there is one parameter to indicate whether the bounding box contains an actual object and is accurately placed. Also, there are three parameters to indicate whether a box belongs to the classes "car," "pedestrian," or "bicycle." 3. Finally, there are the 5 additional parameters used by the region proposal network to estimate accurate object orientations and boundaries. 3D bounding box re-conversion: One of the aspects that makes Complex YOLO special is the extension of the classical Grid RPN approach, which estimates only bounding box location and shape, by an orientation angle, which it encodes as a complex angle in Euler notation (hence the name "E-RPN") such that the orientation may be reconstructed as arctan2(Im,Re)arctan2(Im,Re). The regression values are then directly passed to the computation of a loss function, which is based on the YOLO concept (i.e., the sum of squared errors) using multi-part loss but extends it by an Euler regression part, which is obtained by calculating the difference between the ground truth and the predicted angle that is always assumed to be inside the circle shown above. Why use Complex YOLO? One of the major advantages of the Complex YOLO networks is its speed in comparison to other currently available methods. This makes it well suited for real-time applications such as autonomous vehicles. One disadvantage of the current implementation of Complex YOLO though is the lack of bounding box height and vertical position. All bounding boxes are assumed to be located on the road surface, and height is set to a pre-defined constant based on the detection class. In the tracking stage, this might lead to inaccuracies for driving scenarios with varying elevation. An improved version of Complex YOLO, which extends the concept to full 3D, is described in many papers; rankings and challenges such as the Waymo Open Challenge or the KITTI 3D Object Detection Benchmark and metrics such as "Average Precision" (AP) and "mean Average Precision" (mAP) are often used as the de-facto standard for comparing detection algorithms in a meaningful way. The following figure shows the leaderboard of the Waymo Open Challenge: The board is sorted after the AP/L1 column, which ranks the various algorithms using a single number between 0.0 and 1.0. Object detection algorithms need to perform two main tasks, which are: 1. to decide whether an object exists in the scene and 2. to determine the position, the orientation, and the shape of the object. The first task is called "classification," while the second task is most often referred to as "localization." In real-life scenarios, a scene will consist of several object classes (e.g., vehicles, pedestrians, cyclists, traffic signs), which leads to the necessity of assigning a confidence score to each detected object, usually in the range between 0.0 and 1.0. Based on this score, the detection algorithm can be assessed at various confidence levels, which means that the threshold for accepting or rejecting an object is systematically varied. In order to address these issues, the average precision (AP) metric was proposed as a suitable metric for object detection. To understand the concept behind AP, you first need to understand the terms "precision" and "recall" as well as the four characteristics "true positive" (TP), "true negative" (TN), "false positive" (FP), and "false negative" (FN). TP, TN, FP, and FN: Consider again the point cloud from the last example, now with a set of bounding boxes added by an object detection algorithm: The bounding box labelled "TP" shows a correct detection: It encloses an actual object, and both the shape and the heading are accurate. This is called a "true positive," where "positive" denotes the presence of an object as seen by the detector. On the right, there is a red bounding box labelled "FP," which does not contain any object and thus is a false detection. Such erroneous objects are called "false positive," as the detector wrongly believes in the existence of an actual object. Further, at the bottom of the image, there is an object clearly visible in the point cloud, but the detector has not found it. Such missed detections are called "false negatives," where "negative" means that the detector believes in the absence of an object. In a medicinal trial, where patients are tested for a specific illness, the following would hold (Fig. 1):

TP: Patient actually has the illness, and the test result was positive. FP: Patient does not have the illness, but the test was positive, nonetheless. FN: Patient has the illness, but the test was wrongly negative. Lastly, there is the case where the patient does not have the illness and the test correctly returns a negative result. This is referred to as "true negative" (TN). In object detection, this would mean that there is no object present in a scene, and the detector has correctly not returned a detection.

These four states can be arranged in a matrix, which is called "confusion matrix" in machine learning: Note that the cases, where the predicted values are congruent with the actual values, lie on the main diagonal of the matrix. Precision and recall imagine that we wanted to assess the detection algorithm used in the example

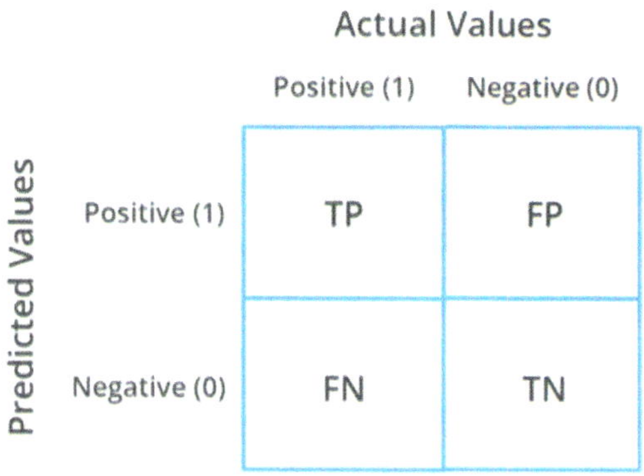

Fig. 1 Confusion matrix

above in a meaningful way. Two questions we might ask could be: 1. What is the probability that an object found by the algorithm actually corresponds to a real object? 2. What is the probability for a real object to be found by the detector? Let us try to answer the first question based on the positives and negatives: In order to arrive at this probability, we need to divide the true positives by the number of all detections, which is the sum of the true positives and the false positives. This ratio is called "precision" P. Obviously, it does not make much sense to compute this metric for a single frame with such a small number of measurements. Hence, in practice, precision and recall are computed over an entire dataset and should contain a number of representatives from each class (TP, TN, FN) such that the result is statistically relevant. To meaningfully assess the performance of a detector, we must examine both precision and recall. A perfect detector would have a recall of 1.0 and a precision of 1.0 at the same time. In practice however, both measures are often in tension, which means that improving precision, e.g., by adapting some detector parameters, will most probably reduce recall and vice versa. Assume that we have a detector, which returns the following results: As you can see, precision has decreased, while recall has increased. In case of an increased classification threshold, this effect would reverse. In general, a detection algorithm that outperforms another algorithm on both precision and recall is most likely better. Another measure that is also based on the prediction types is called "accuracy," which is the ratio of the number of correct predictions and the total number of predictions. Accuracy can be expressed as: In object detection though, accuracy is not used as the number of true negatives does not correspond to a meaningful detector behavior. The state "there is no object and the detector has not detected one" would hold for all areas of a scene without detection and without objects. In practice, we require the presence of an object, be it from a detector or from a trusted source such as a human observer, to derive the states TP, FP, and FN. Intersection-over-Union (IoU): In the previous examples, we have taken for granted that some entity would provide us with the information whether a detection was TP, FP, or FN (Fig. 2).

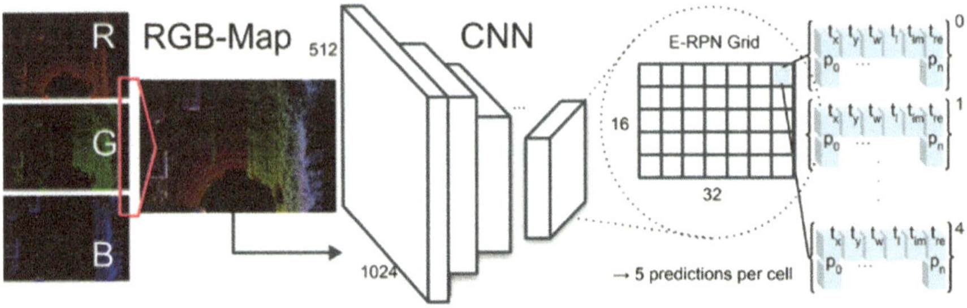

Fig. 2 Complex RS-ResNet network architecture

In order to classify detections into these categories, several steps need to be taken: 1. Provide a list of human-generated detections, which serve as ground truth. Such detections are often referred to as "labels." 2. Match the list of detections provided by the algorithm that we want to evaluate with the list of ground-truth labels. 3. Decide whether a certain detection is TP, FP, or FN based on the degree of overlap between detection and ground-truth label. Take a look at the point cloud from the previous example, which now contains bounding boxes generated by the detector (red) as well as ground-truth labels (green). In order to assess the detector in a meaningful way though, it is required to decide whether a ground-truth label and a detection result are treated as a match—or not. In the final assignment, measuring IoU "Intersection-over-Union" (IoU) is used to decide whether to assign two bounding boxes. The idea is to measure the degree of overlap between each bounding box returned by the detector and all ground-truth bounding boxes in such a way that the area of intersection and the area of the union of both rectangles are computed. The following figure illustrates the concept, the smaller the overlap between detection and ground truth, the smaller will be the intersection-over-union. In case, there is no overlap at all, the IoU will be at 0.0, while for a perfect match, the IoU will be 1.0. In practice, we need to decide on an "appropriate" threshold for the IoU when matching detections and ground-truth labels. For example, if the IoU threshold is 0.5 and the IoU value for a certain label-detection match is 0.7, then we classify it as TP. On the other hand, when the IoU is below the threshold, e.g., at 0.3, we classify the corresponding detection as FP. In case there are ground-truth labels, for which there are no detections at all, these are classified as FN.

4 Results and Performance Analysis

Based on the IoU threshold, the values for precision and recall will change, since at a low threshold, it is expected that many detections will be matched with labels, resulting in a high number of TP. In most benchmarks (e.g., KITTI, Waymo), the required minimum IoU is 70% for vehicles. Precision–Recall Curve: As mentioned previously, there is an inverse relationship between precision and recall, which is dependent on the threshold we choose for object detection. Let us take a look at the following figure, which shows the precision–recall curves of various object detection algorithms (each color represents a different algorithm): The curves are generated by varying the detection threshold from 0.0 to 1.0 and by computing precision and recall for each setting. Based on this metric, the solid blue curve shows the best performance, as the precision (i.e., the likelihood that a detection actually corresponds to a real object) drops the least for increasing recall values. Average Precision: The idea of the average precision (AP) metric is to compact the information within the precision–recall curve into a single number, which can be used to easily compare algorithms with each other. This goal is achieved by summing the precision values for different (= 11) equally spaced recall values: But varying the threshold level in equally spaced increments does not correspond to equally spaced increases in recall (Figs. 3 and 4).

The AP score for an algorithm varies between 0.0 and 1.0 with the latter being a perfect result. Mean Average Precision (mAP): Based on the observation that changing the IoU threshold affects both precision and recall, the idea of the mean average precision (mAP) measure is to compute the AP scores for various IoU thresholds and then the mean from these values. The following figure shows precision–recall curves for several settings of the IoU threshold: As can be seen, the shape of the curve approaches a rectangle for decreasing IoU thresholds. Also, some ranking sites not only consider IoU thresholds but various object classes as well. In the PASCAL VOC 2007 challenge for example, only a single IoU setting, (0.5) and 20 object classes were considered. In the COCO 2017 challenge on the other hand,

Leaderboard

Method Name	Object Type	Sensors	Frames [-p, +f]	AP / L1	APH / L1	AP / L2	APH / L2	Date
	VEHICLE	L						
HorizonLiDAR3D	VEHICLE	CL	[-4, +0]	0.8509	0.8468	0.7823	0.7783	2020-05-31
PV-RCNN	VEHICLE	L	[-1, +0]	0.8106	0.8057	0.7369	0.7323	2020-06-01
TS-LidarDet	VEHICLE	L	[-0, +0]	0.8075	0.8018	0.7265	0.7212	2020-06-01
AMPP-Noah	VEHICLE	L	[-0, +0]	0.7929	0.7887	0.7085	0.7047	2020-05-31
SJTU AI Institute	VEHICLE	CL	[-0, +0]	0.7554	0.7510	0.6834	0.6793	2020-12-06
Simple Baseline v2	VEHICLE	L	[-3, +0]	0.7508	0.7452	0.6640	0.6591	2020-06-01
Det3D-Waymo-3D-FS-VS	VEHICLE	L	[-0, +0]	0.7367	0.7269	0.6603	0.6511	2020-05-31
mxlidar	VEHICLE	L	[-0, +0]	0.7358	0.7309	0.6490	0.6446	2020-06-01
Det3D-Waymo-3D	VEHICLE	L	[-0, +0]	0.7329	0.7227	0.6521	0.6429	2020-05-29
RCD_RCNN	VEHICLE	LT	[-0, +0]	0.7197	0.7159	0.6506	0.6470	2020-05-24

Waymo Challenge Leaderboard

Fig. 3 Waymo competition leaderboard

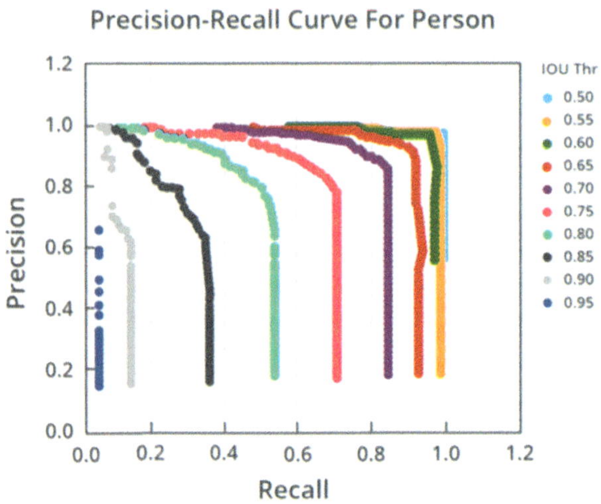

Fig. 4 Confusion matrix for the predicted object classes

the mAP was averaged over 10 IoU settings and 80 object classes. In the Waymo 2021, challenge has two levels of labels hierarchy, 4 object classes, and three levels of lidar range images with the associated camera images are given, 3D label boxes need to be predicted for the objects in the image.

The efficient model is defined by the Waymo 2021 challenge as a model running at 70 ms/frame or faster on a Nvidia Tesla V100 GPU needs to be considered and also with APH/L2 should be greater than 70 in order to be given considered for top position in leaderboard. Our model achieves the running time at 70 ms/frame and faster rates in more powerful GPU as the RS-ResNet architecture is scalable to new GPU architectures easily. The threshold value of 70ms is calculated based on the scan frequency of a Lidar unit, which stands at 10HZ, and 70ms is the time interval given to the algorithm to detect and classify objects.

5 Conclusion

With latest RS-ResNet family of re-scaled convolution networks, the task of 3D object detection can be performed with higher speed and better detection prediction with point clouds having varying degrees of visibility. Geometrical overlap of predicted labels and ground truths is identified for each prediction and compared with the state-of-the-art architectures. The future scope of the work involves identifying the motion characteristics of the objects detected using algorithms such as Kalman Filter and Extended Kalman Filter in the detection pipeline, which in turn makes autonomous driving of a vehicle more ready for unconstrained environments such as Indian roads. And many emerging methodologies in the field follow the similar approach of 3D object detection with velocity estimation for localization and path planning of an autonomous vehicle, and the proposed methodology helps in progressing the field by making the model hardware independent and less hardware resource-intensive.

References

1. A. Bochkovskiy, C.-Y. Wang, H.-Y.M. Liao, Yolov4: optimal speed and accuracy of object detection. Preprint. arXiv:2004.10934 (2020)
2. X. Chen, A. Gupta, An implementation of faster RCNN with study for region sampling. Preprint. arXiv:1702.02138 (2017)
3. R. Girshick, Fast R-CNN, in *Proceedings of the IEEE International Conference on Computer Vision* (2015), pp. 1440–1448
4. K. He, G. Gkioxari, P. Dollár, R. Girshick, Mask R-CNN, in *Proceedings of the IEEE International Conference on Computer Vision* (2017), pp. 2961–2969
5. X. Long, K. Deng, G. Wang, Y. Zhang, Q. Dang, Y. Gao, H. Shen, J. Ren, S. Han, E. Ding, et al., PP-YOLO: an effective and efficient implementation of object detector. Preprint. arXiv:2007.12099 (2020)
6. C.R. Qi, H. Su, K. Mo, L.J. Guibas, PointNet: deep learning on point sets for 3D classification and segmentation, in *Proceedings of the IEEE Conference on Computer Vision and Pattern Recognition* (2017), pp. 652–660
7. S. Ren, K. He, R. Girshick, J. Sun, Faster r-CNN: towards real-time object detection with region proposal networks. Adv. Neural Inf. Proces. Syst. **28**, 91–99 (2015)
8. X. Wang, A. Shrivastava, A. Gupta, A-fast-RCNN: hard positive generation via adversary for object detection, in *Proceedings of the IEEE Conference on Computer Vision and Pattern Recognition* (2017), pp. 2606–2615
9. C.-Y. Wang, A. Bochkovskiy, H.-Y.M. Liao, Scaled-YOLOv4: scaling cross stage partial network, in *Proceedings of the IEEE/CVF Conference on Computer Vision and Pattern Recognition* (2021), pp. 13029–13038
10. D. Wu, S. Lv, M. Jiang, H. Song, Using channel pruning-based YOLO v4 deep learning algorithm for the real-time and accurate detection of apple flowers in natural environments. Comput. Electron. Agric. **178**, 105742 (2020)

Part III
Bigdata Emerging Applications

Various Physiological Methods to Identify Sleep Onset

Cheng-You Gan, Ting-Hsun Huang, Kuan-Min Huang, and Chan-Yun Yang

Abstract Sleep occupies a large part of human life. Good sleep has a positive impact on physical health. For people who have difficulty in sleeping, or have even become sick, it is very important to have a device to help sleep. And a mechanism for judging sleep onset is very important for the development of sleep aids. This article lists three common and effective methods to detect sleep episodes: (1) electrocardiography (ECG), (2) electroencephalography (EEG), and (3) electrooculogram (EOG). The focus of this article is to compare the differences and success rates of these three detection methods, hoping to find a method that is most suitable for the development of sleep aids.

Keywords Sleep onset · ECG · EEG · EOG

1 Introduction

Sleep quality is the assessment of how well the subject's sleeping, i.e., the degree to which restful sleep is sustained during the sleeping time and the subject feels fresh on waking and throughout the workday. Conventionally, the assessment of sleep quality includes synthetically latency until sleep onset, wakefulness after sleep onset, and duration of the sleep [1]. Here, the sleep onset as defined as the "transition from waking to sleep" is obviously a key for the assessment. Generally, we often divide the stages of sleep into three stages: awake, non-rapid eye movement (NREM) sleep, and rapid eye movement (REM) sleep, where NREM can be divided into: The first stage: occurs at the beginning of sleep, when the eyeballs move slowly, the alpha wave disappears, and theta wave appears. At this stage people still think that they are awake, but feel drowsy. The second stage: This stage people enter the unconscious stage, but it is easy to wake up. The eyeballs no longer

C.-Y. Gan · T.-H. Huang · K.-M. Huang · C.-Y. Yang (✉)
National Taipei University, Electrical Engineering, New Taipei, Taiwan
e-mail: cyyang@mail.ntpu.edu.tw

A. Haldorai et al. (eds.), *4th EAI International Conference on Big Data Innovation for Sustainable Cognitive Computing*, EAI/Springer Innovations in Communication and Computing, https://doi.org/10.1007/978-3-031-07654-1_8

move, nor do they dream too much. In the EEG records at this time, there will occasionally be faster sleep spindle waves and K-complex waves. The third stage: This stage is the deepest sleep stage in NREM, also known as slow-wave sleep. At this time, the delta wave, which represents deep sleep, will dominate. According to the recommendations of the National Sleep Foundation, the sleep time is 14–17 hours recommended for newborns, 12–5 hours for babies, 11–14 hours for toddlers, 10–13 hours for preschoolers, and 9–11 hours for school-aged children. It's also recommended that teenagers should have 8–10 hours of sleep, young adults and adults should get 7–9 hours, and older persons should get 7–8 hours. If the sleep time is less the recommended hours, it would be harmful to health [2]. According to the research of Gottlieb et al., the sleep time is also related to diabetes mellitus and impaired glucose tolerance [3]. It can be seen that sleep has a great relationship with physical health. As a pre-study to discover the fact of sleep onset, we make a detailed survey to discuss sleep onset with three common measurement methods.

2 Methods of Measuring Sleep Onset

2.1 ECG

Electrocardiography (ECG) is a type of electrophysiological activity that uses electrodes to contact the skin of a body and records the heart's electrophysiological activity in units of time. The change is measured by the recording device and zoomed in to draw the ECG. Table 1 lists kinds of the wave patterns that can be observed in ECG. Figure 1 shows the wave pattern that can be observed on the ECG during a normal sinus rhythm.

In the general cases, the heart rate is in the midst of the baroreflex control of blood pressure since a decreased heart rate causes blood pressure to rise. During sleep, this baroreflex regulation is more active in order to keep blood pressure low [6, 7]. However, Park et al. have studied the difference among 10 subjects and estimated the error in the work of [8]. According to their results, the judgment recommended to the professionals is 0.25 ± 0.35 minutes.

One the other parameter commonly used for the sleep onset detection in ECG is heart rhythm variability (HRV), which is defined as the change in the time interval between two consecutive heartbeats. As found by Szypulska and Piotrowski [9], there is a correlation between sleep onset and HRV parameters. The lower the HRV parameter, the more regular the heartbeat, which can be used as a basis for the onset of sleep.

Table 1 Common waveforms in ECG [4]

Feature	Description	Duration
P wave	Depolarization of the atria is represented by the P wave. From the SA node to the AV node, and from the right atrium to the left atrium, atrial depolarization occurs	<80 ms
PR interval	The PR interval is the distance between the start of the P wave and the start of the QRS complex. The time it takes for an electrical impulse to go from the sinus node to the AV node is reflected in this interval	120–200 ms
QRS complex	The rapid depolarization of the right and left ventricles is represented by the QRS complex. The QRS complex has a much bigger amplitude than the P wave because the ventricles have a much larger muscular mass than the atria	80–100 ms
J-point	The J-point marks the end of the QRS complex and the start of the ST segment	
ST segment	The ST segment is the period when the ventricles are depolarized and connects the QRS complex and the T wave	
T wave	The repolarization of the ventricles is represented by the T wave. Except for aVR and lead V1, it is normally upright in all leads	160 ms
Corrected QT interval (QTc)	The QT interval is the distance between the start of the QRS complex and the end of the T wave. Acceptable ranges vary depending on heart rate, therefore divide by the square root of the RR interval to get the QTc	<440 ms
U wave	The repolarization of the interventricular septum is thought to be the cause of the U wave. It usually has a small amplitude and, more often than not, is completely absent	

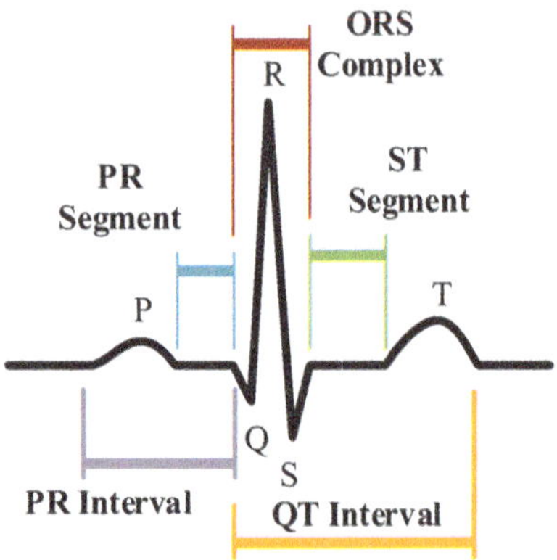

Fig. 1 ECG of a heart with normal sinus rhythm [5]

Table 2 General classifications of EEG bands

Band	Frequency (Hz)
Delta	<4 Hz
Theta	4–7 Hz
Alpha	8–15 Hz
Beta	16–31 Hz
Gamma	>32 Hz

2.2 EEG

Electroencephalography (EEG) is a technique for recording brain waves. When multiple pyramidal cells are activated, electrodes are placed on the scalp to record the voltage fluctuations created by the ionic current of the brain neurons, and what is recorded is the synchronous sum of the postsynaptic potentials [10]. In general, the brain waves of EEG spectrum can be decomposed as many frequency bands as those described in Table 2.

In the recent decades, more and more researchers use EEG to diagnose sleep stages or sleep episodes. Some of them use kinds of neural network to extract features and identify the stages [11–13] where the brain waves are staged into delta, theta, alpha, and beta bands by calculating their corresponding power spectrum and therefore the identification could be performed on the task [14–17]. Chen et al. [13] discovered first that the alpha band is mainly found when people are awake, and when people enter sleep stage 1, theta wave will become the main band. Alternatively, Feng and Gu in their work [14] also observed that alpha power occurs when people entering sleep. The phenomenon of falling and rising theta power together with the discovery of [13] have led to a development of a sleeping aid device, which successfully shortened the time for people to fall asleep.

For studying sleep onset, one another interesting way is through entropy, which is defined as the irregularity of the EEG wave pattern. The irregularity measures the brain activities with a kind of correlations. The higher the uncertainty (irregularity), the higher the value of entropy. In previous studies, researchers found that the EEG frequency of people declined after they enter sleep. The wave pattern is gradually decreased [11–17]. It means that the entropy could be admissible for judging whether humans are going to sleep or not [18–21]. In the study of Mu et al. [21], four types of common entropies were used to extract feature for detecting driver fatigue where EEG signals were used as the signal source of the defined entropies. The aggregation of the entropies tackled the complexity in managing the mixture of the EEG bands and reduced significantly the difficulty of signal decomposition and identification. Experimental results showed that, for the 12 subjects, the highest classification rate reached 98.75%.

2.3 EOG

Electrooculogram (EOG) measures the electrostatic potential of the retina that exists between the retinal pigment epithelium and photoreceptor cells. The electrical response of the retina comes from the complex structure of the retinal nerve network. The photoreceptor epithelium of the retina has a positive potential, and the direction of the pigment epithelium is a negative potential. The potential difference between the two layers can reach 60 mV. The potential recorded varying time forms EOG [22].

A general EOG measurement installed an electrode at 1 cm above the outer corner of the left eye (E1) and one the other reference electrode on the left mastoid (A1) to obtain the EOG trace output. In general, EOG signals are expressed in the edf format. In the work of [23], Huang et al. employed the EOG measurement to trace blinking and slow eye movements (SEM) of the subject's eyes and used them as the criteria for sleep onset assessment. Their research reported the consistency with the EEG results reached up to 91.16% with a sensitivity 93.33%. A similar method was introduced by Magosso et al. [24] with a little architecture change. In the work of Magosso et al., the number of EOG channels was increased to two, and wavelet transform signal processing was performed additionally. The outcome arrived 88% agreement on the sleep onset assessment with 81% sensitivity and 63% specificity.

3 Discussions

At present, the main equipment for measuring physiological signals in the hospital is polysomnography (PSG), which is a widely used equipment for patients undergoing multiple channels physical examination of sleep. In general, it is necessary to configure the multiple channel device on the patient to stay in the sleep center for 6–8 hours. Multiple electrode patches are attached to measure EEG signals from the patient, which provide the sleep specialist a reference basis for distinguishing sleeping stages. Together with the EEG patches, 2–3 ECG electrode patches are

Table 3 Comparison of EEG, ECG, and EOG signals in polysomnography

	ECG	EEG	EOG
Electrode patch position	Above the ribs	Scattered on the skull	Around the eyes
The number of common electrode patches in clinical practice	12	>20	4
Judgment basis for sleep onset	Heart rate baroreflex regulation HRV	Frequency entropy	Blink SEM

placed beside the patient's ribs to record various ECG signals, and 2 EOG electrode patches attached to the outer corners of both the patient's eyes to catch the EOG signals which is the potential difference between the cornea and the retina. Table 3 shows the comparison of three physiological signals in PSG.

Beside the EEG, ECG, and EOG measurements, there are more other physiological signals which can be obtained from PSG. Since the exceptional signals are not the focused concern of this sleep onset study, they can be ignored from the paper.

If the physiological parameters between wakefulness and sleep can be extracted, the sleep onset could be observed by a decrease in EEG alpha band power and an increase in its delta band power. Instead of the examination that no blinking signal is generated in EOG, an alternative SEM examination from EOG could be also used to detect the drowsing. It was also found in ECG that the heartbeat after entering sleep was slower than that when waking up. Although the above physiological signals are gradually adopted with their reliable accuracies and consistency, there are still some problems that still need to be solved. All physiological detection devices need to be pasted or even invasive to the patient's body to avoid noisy disturbance and make the collected signals better. Close to the real situation, this way of sticking or intruding will increase the user's discomfort to some extent, which is contrary to the original intention of developing the sleep aid device.

The research purpose of sleep aids is to improve sleep quality by reducing the time to get sleep. If people enter sleep, various sleep-help interventions (e.g., music) may affect sleep quality, in order to avoid this. The occurrence of a good wakefulness and sleep discrimination is crucial.

4 Conclusion

This survey paper discusses the methods of sleep onset detection with the three common physiological signals of ECG, EEG, and EOG. These three measurements are currently major and straightforward methodological sources for the detection. Although there are still some other physiological signals suggested by the researchers, the straightforward potential makes most of the developments inclined to the adoption of ECG, EEG, and EOG sources. So, this article surveyed specifically only the above three common physiological signals for the purpose of sleep onset detection. The survey would help successive researches to discover the relevant physiological parameters to engage the detection. According to the survey, all three physiological signals are agreed in the use to study sleep disorders well.

References

1. M. Hirshkowitz et al., The effects of medications on sleep quality and sleep architecture – Up-to-date. Sleep Health **1**, 40–43 (2015) https://www.uptodate.com/contents/the-effects-of-medications-on-sleep-quality-and-sleep-architecture
2. Naming of the waves in the ECG, with a brief account of their genesis. Hurst **98**(18): 1937 – Circulation. https://web.archive.org/web/20100520011550/http://www.circ.ahajournals.org/cgi/content/full/98/18/1937
3. D.J. Gottlieb et al., Association of sleep time with diabetes mellitus and impaired glucose tolerance. Arch. Intern. Med. **165**, 863–867 (2005)
4. D.W. Jung et al., Nocturnal awakening and sleep efficiency estimation using unobtrusively measured ballistocardiogram. I.E.E.E. Trans. Biomed. Eng. **61**, 131–138 (2014)
5. Electrocardiography. BioNinja. https://ib.bioninja.com.au/standard-level/topic-6-human-physiology/62-the-blood-system/electrocardiography.html
6. M. Ursino, E. Magosso, Role of short-term cardiovascular regulation in heart period variability: A modeling study. Am. J. Phys. Heart Circ. Phys. **284**, H1479–H1493 (2003)
7. J.S. Loredo, R. Nelesen, S. Ancoli-Israel, J.E. Dimsdale, Sleep quality and blood pressure dipping in normal adults. Sleep **27**, 7 (2004)
8. K.S. Park, S.H. Hwang, D.W. Jung, H.N. Yoon, W.K. Lee, Ballistocardiography for nonintrusive sleep structure estimation, in *2014 36th Annual International Conference of the IEEE Engineering in Medicine and Biology Society* (IEEE, 2014), pp. 5184–5187. https://doi.org/10.1109/EMBC.2014.6944793
9. M. Szypulska, Z. Piotrowski, Prediction of fatigue and sleep onset using HRV analysis, in *Proceedings of the 19th International Conference Mixed Design of Integrated Circuits and Systems – MIXDES 2012*, (2012), pp. 543–546
10. J.C. Henry, Electroencephalography: Basic principles, clinical applications, and related fields, Fifth Edition. Neurology **67**, 2092–2092 (2006)
11. T. Shimada, T. Shiina, Y. Saito, Detection of characteristic waves of sleep EEG by neural network analysis. I.E.E.E. Trans. Biomed. Eng. **47**, 369–379 (2000)
12. L.-J. Kau, K.-M. Chao, A Takagi-Sugeno fuzzy neural network-based algorithm with single-channel EEG signal for the discrimination between light and deep sleep stages, in *2016 IEEE Biomedical Circuits and Systems Conference (BioCAS)*, (IEEE, 2016), pp. 532–535. 10.1109/BioCAS.2016.7833849
13. X. Chen, B. Wang, X. Wang, Automatic sleep stage classification for daytime nap based on hopfield neural network, in *2013 25th Chinese Control and Decision Conference (CCDC)*, (IEEE, 2013), pp. 2671–2674. 10.1109/CCDC.2013.6561393
14. Z. Feng, F. Gu, Power spectral analysis of recovery sleep of sleep deprivation and hypnotic drug induced sleep, in *2005 IEEE Engineering in Medicine and Biology 27th Annual Conference*, (IEEE, 2005), pp. 3593–3596. https://doi.org/10.1109/IEMBS.2005.1617258
15. Z. Zhang et al., Reduction in time-to-sleep through EEG based brain state detection and audio stimulation, in *2015 37th Annual International Conference of the IEEE Engineering in Medicine and Biology Society (EMBC)*, (IEEE, 2015), pp. 8050–8053. 10.1109/EMBC.2015.7320261

16. Z. Zhang et al., Automatic sleep onset detection using single EEG sensor, in *2014 36th Annual International Conference of the IEEE Engineering in Medicine and Biology Society*, (IEEE, 2014), pp. 2265–2268. https://doi.org/10.1109/EMBC.2014.6944071
17. A. Germain, EEG power associated with early sleep onset images differing in sensory content. Sleep Res. Online **4**, 83–90 (2001)
18. T. Zobaed et al., Real time sleep onset detection from single channel EEG signal using block sample entropy. IOP Conf. Ser.: Mater. Sci. Eng. **928**, 032021 (2020)
19. N. Sriraam, P.T.K. Shri, U. Maheshwari, Recognition of wake-sleep stage 1 multichannel eeg patterns using spectral entropy features for drowsiness detection. Australas. Phys. Eng. Sci. Med. **39**, 797–806 (2016)
20. C. Bandt, A new kind of permutation entropy used to classify sleep stages from invisible EEG microstructure. Entropy **19**, 197 (2017)
21. Z. Mu, J. Hu, J. Min, Driver fatigue detection system using electroencephalography signals based on combined entropy features. Appl. Sci. **7**, 150 (2017)
22. Electrooculogram – EyeWiki. https://eyewiki.aao.org/Electrooculogram
23. R.-J. Huang et al., Strong correlation of sleep onset between EOG and EEG sleep stage 1 and 2, in *2012 International Symposium on Computer, Consumer and Control*, (IEEE, 2012), pp. 614–617. https://doi.org/10.1109/IS3C.2012.160
24. E. Magosso, M. Ursino, F. Provini, P. Montagna, Wavelet analysis of electroencephalographic and electro-oculographic changes during the sleep onset period, in *2007 29th Annual International Conference of the IEEE Engineering in Medicine and Biology Society*, (IEEE, 2007), pp. 4006–4010. https://doi.org/10.1109/IEMBS.2007.4353212

Internet of Things (IoT) Adoption: Challenges Among Small and Medium Scale Enterprises (SMEs)

Praveen M. Kulkarni , Prayag Gokhale , and Padma S. Dandannavar

Abstract Modern production offers great prospects to capitalize on the latest product development technologies. Innovative technologies like the Internet of Things improve the information process and improve productivity in production. To gain a deeper understanding of IoT implementation in small and medium enterprises, this study was conducted to understand the relationship between the Internet of Things and its use in small and medium enterprises. The research has also considered the role of management skills required for the effective use of the Internet of Things for small and medium enterprises. The results showed that the role of management skills and transparency concerning IoT implementation objectives is critical for effective implementation in small and medium enterprises.

Keywords Internet of things · Technology management · Technology adoption

1 Introduction

Smart manufacturing is influencing the manufacturing process wherein a substantial number of machines are networked for the manufacturing process [1]. The networked machines have dedicated activities and are interconnected with other machines for production output. The connected machines are assembled to enhance flexibility for customized production processes [2]. Hence, smart manufacturing

P. M. Kulkarni (✉)
K.L.S. Gogte Institute of Technology, Department of MBA, Belagavi, Karnataka, India

P. Gokhale
KLE Dr. M. S. Sheshgiri College of Engineering and Technology, Department of MBA, Udyambag Belgaum, Karnataka, India

P. S. Dandannavar
K.L.S. Gogte Institute of Technology, Department of Computer Science Engineering, Belagavi, Karnataka, India
e-mail: padmad@git.edu

A. Haldorai et al. (eds.), *4th EAI International Conference on Big Data Innovation for Sustainable Cognitive Computing*, EAI/Springer Innovations in Communication and Computing, https://doi.org/10.1007/978-3-031-07654-1_9

supports improving the performance of manufacturing. Smart manufacturing is enabled with the technology of IoT; this technology is developed with large information networks which are interconnected "things," comprising of machinery, sensors, actuators, regulators, automatons, and workforces [3]. The manufacturing process is well integrated through these "things" through the Internet-based Internet of Things (IoT) platform of production. The links between production and materials are linked to new resource planning, supply chain management, production system, and process management system [3–6].

Recognizing the full potential of IoT requires a combination of management skills and production capabilities. One of the most common factors that contribute to the achievement of management skills and the ability to generate investment in these technologies is SMEs [7, 10, 11].

Therefore, the present study investigates managerial and manufacturing capabilities for the successful implementation of IoT in SMEs. Further, studies related to IoT in SMEs have been undertaken from the perspective of vision and challenges for realizing the Internet of Things [8] focused on overall manufacturing and IoT [2]. This shows that there are very few studies that have the combination of IoT capabilities and managerial capabilities and mediating factors for Internet of Things SMEs. Therefore, this study presents the challenges with regard to managerial skills and implementation of IoT and presents the study results with regard to objectives of implementation and managerial capabilities in SMEs.

2 Literature Review

This section includes the reviews of the various works of literature on the Internet of Things and the role of managerial skills; this section also includes factors influencing the implementation of the Internet of Things in SMEs.

2.1 Application of IoT

In the manufacturing process, application of IoT has been applied to gain an advantage and become competitive in manufacturing. Ren et al. [13] mention that IoT provides real-time information on manufacturing, which improves supply chain and manufacturing in SMEs. Further, IoT indicates focused operational performance for the success of manufacturing [13]. The study conducted by Denkena et al. mentioned that SMEs need to apply RFID technology with IoT for the flow of information from a machine for effective decision-making on manufacturing [14]. This technology is applied in information related to production machines which support continuous improvement in manufacturing [12]. The data generated through information flow is a large number; employees need to be trained in handling the information and converting it into meaningful information of production; this

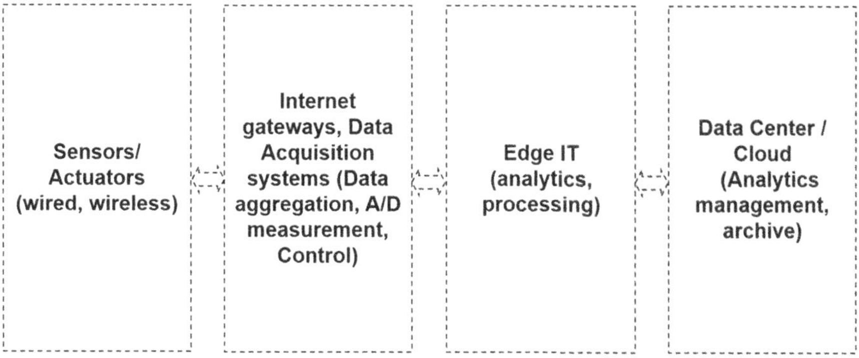

Fig. 1 IoT architecture

was achieved through training the employees on IoT [9]. Most of the research related to IoT is focused on improving automation and flexibility in manufacturing [15]. The above findings with regard to the application of IoT are related to enhancing the production process, quality management, and faster information flow of manufacturing.

2.1.1 IoT Architecture Used for the Research

Figure 1 shows IoT architecture adopted and used during the research work. The IoT architecture depicts the various steps that are followed during the implementation and adoption of IoT in various SMEs. The IoT architecture uses these steps and analyze the various managerial capabilities at different stages of the IoT process and analyzes the relationship between the various managerial capabilities on performance. The research works also focus on the important components of IoT like sensors/devices, connectivity, data processing, and user interface.

2.2 *Objectives of Implementation of IoT*

Manufacturing SMEs implement IoT with specific objectives to achieve; these objectives include (a) flexibility in manufacturing, (b) cost reduction in manufacturing, (c) improving productivity in manufacturing, (d) enhancing quality, and (e) reducing time of delivery during manufacturing [19, 20]. The flexibility objective is focused on responding to the flow of information in real time; application of manufacturing planning through cloud computing enhances the productivity and flexibility in the factory [16]. A second goal is to reduce production costs through IoT; the use of digital tools has been supported to reduce production costs through this technology [17]. Thirdly, enhancing the production from data formatting and sequencing improvement of machines was possible through the flow of information through IoT [13]. Product quality is improved through the application of the RFID part to the total manufacturing process; that is why IoT plays an important role in improving product quality [18]. IoT use in reduced production improves production flow and improves delivery during production [14].

2.3 Industrial Managerial Capacity

Management skills are essential to the successful use of technology in production. Alexandre M et al. provided the framework of management skills for understanding Industrial 4.0, which also integrates with IoT; the elements related to management capabilities are (i) control, (ii) monitoring, (iii) independence, and (iv) efficiency. The monitoring factor in the production process is enhanced by monitoring capabilities. The IoT uses sensors in the production process to monitor the flow of information and provides warnings on information blocking; this enhances the production process in SMEs [20]. This flow of information is achieved through the use of embedded RFID technology, which facilitates the recording of information and improves production recognition. The second variable is controlled as a controlled force, which provides past production data and reflects the level of acceptance in production [18]. The use of RFID technology in production allows comparisons with previous data and brings better control measures in production. The following variations in management skills include performance improvement focused on improving processes and achieving a high level of productivity; results concerning SMEs have shown positive relationships in terms of productivity and IoT. Finally, variability in management skills is related to autonomy, which indicates that IoT implementation requires a variety of strategies and technologies in order to effectively use these technologies [16, 21–24]. The above literature on IoT use, IoT adoption goals, and the role of management skills for IoT use highlighted these questions in the research below:

Q1: The relationship between IoT use and IoT management capacity in SME production

Q2: The role of management skills in achieving IoT adoption goals in producing SMEs

The discussion above in the literature review complements this research novel as it demonstrates the combination of IoT use in objectives of IoT and the role of management skills in IoT adoption in SME production.

Table 1 Summary of profile of the participants

Sr. No.	Respondents Profile	Number	Percentage
1	100–150 employees	15	68.18
2	70–100 employees	3	13.64
3	50–70 employees	2	9.09
4	Less than 50	2	9.09
	Total	22	100

3 Research Methods

The study research design provides a roadmap for the collection of data and measurement of data. In this study, the exploratory research design is applied to understand the adoption of IoT in SME manufacturing. Data were obtained through a systematic questionnaire, and data was collected from SMEs used by IoT in SMEs in Belagavi, Karnataka, India. Collected data is analyzed by a paired t-test sample that determines the variance among the viewing set. Therefore, in this study, the difference between the use of IoT in IoT implementation objectives and management capacity and IoT implementation objectives is studied. The use of a paired t-test sample is appropriate as data are collected on a Likert scale in SMEs using IoT in production [25]. The summary of the profile of respondents is presented in the table. Table 1 shows respondents profile which is established based on the number of employees employed in the organizations.

4 Results

The results of the research are presented in this section. The results are presented in Tables 2 and 3. Outcome analysis was presented regarding the relationship between IoT use and management capabilities required for effective IoT use and secondly for management skills and IoT implementation goals.

Effective performance in IoT production and relationship with operational performance showed positive relationship with autonomy 0.02 > 0.005, optimization operation 0.01 > 0.005, and control operational 0.01 > 0.005. However, results with regard to operational performance and monitoring showed negative relationship with 0.013 < 0.005 level of significance. The flow of information has shown a positive relationship and monitoring. Results regarding IoT information flow and monitoring showed positive results with Sig. 0.02 > 0.005 value of value. However, the effects on autonomy were not significantly associated with Sig. 0.08. But in terms of the flow of knowledge and optimization, it has shown a good relationship with Sig. value 0.002 > 0.005 value of value. The fast evaluation showed a positive relationship regarding control, optimization, and autonomy with Sig. value of 0.02, 0.03, and 0.04 > 0.005 value. However, it has shown negative relationship with monitoring significance value of 0.10 value <0.005. The use of IoT

Table 2 Results of application of IoT and managerial capability

Pairs	Relationship	Mean	σ	ϵ	95% level of confidence		t	DF	Sig. (2-tailed)
					Lower	Upper			
I	Operational performance – monitoring	0.59	2.06	0.24	0.11	1.07	2.44	72	0.13
II	Operational performance – control	0.01	1.23	0.14	−0.27	0.30	0.10	72	0.01
III	Operational performance – optimization	0.08	1.15	0.13	−0.19	0.35	0.61	72	0.01
IV	Operational performance – autonomy	0.32	1.47	0.17	−0.03	0.66	1.83	72	0.02
V	Flow of information – monitoring	0.67	1.68	0.20	0.28	1.06	3.42	72	0.02
VI	Flow of information – MC control	0.10	1.16	0.14	−0.17	0.37	0.71	72	0.01
VII	Flow of information – MC optimization	0.16	1.51	0.18	−0.19	0.52	0.93	72	0.02
VIII	Flow of information – autonomy	0.40	1.33	0.16	0.09	0.71	2.55	72	0.08
IX	Fast evaluation of performance – monitoring	−0.73	2.26	0.26	−1.25	−0.20	−2.75	72	0.10
X	Fast evaluation of performance – control	−1.30	1.69	0.20	−1.70	−0.91	−6.58	72	0.02
XI	Fast evaluation of performance – optimization	−1.23	1.85	0.22	−1.67	−0.80	−5.69	72	0.03
XII	Fast evaluation of performance – autonomy	−1.00	2.00	0.23	−1.47	−0.53	−4.27	72	0.04
XIII	Decision-making – monitoring	1.29	1.71	0.20	0.89	1.69	6.43	72	0.11
XIV	Decision-making – control	0.71	1.07	0.13	0.46	0.96	5.67	72	0.01
XV	Decision-making – optimization	0.78	1.12	0.13	0.52	1.04	5.95	72	0.02
XVI	Decision-making – autonomy	1.01	1.27	0.15	0.72	1.31	6.79	72	0.01
XVII	Integration of work – Monitoring	0.84	1.97	0.23	0.38	1.30	3.62	72	0.01
XVIII	Integration of work – control	0.26	1.57	0.18	−0.11	0.63	1.41	72	0.71
XIX	Integration of work – optimization	0.33	1.39	0.16	0.00	0.65	2.01	72	0.01
XX	Integration of work – autonomy	0.56	1.55	0.18	0.20	0.92	3.09	72	0.01

Table 3 Results of managerial capability and objectives of the implementation of IoT

Pairs	Relationship	I	σ	ϵ	95% level of confidence		t	DF	Sig. (2-tailed)
					Lower	Upper			
I	Monitoring – flexibility	−0.41	1.82	0.21	−0.84	0.01	−1.92	72	0.25
II	Monitoring – cost reduction	0.01	1.50	0.18	−0.34	0.36	0.08	72	0.28
III	Monitoring – productivity	−0.96	1.74	0.20	−1.36	−0.55	−4.72	72	0.01
IV	Monitoring – quality	−0.37	1.50	0.18	−0.72	−0.02	−2.10	72	0.20
V	Monitoring – delivery time	0.53	1.55	0.18	0.17	0.90	2.95	72	0.41
VI	Control – flexibility	0.16	1.52	0.18	−0.19	0.52	0.92	72	0.31
VII	Control – cost reduction	0.59	1.54	0.18	0.23	0.95	3.28	72	0.01
VIII	Control – productivity	−0.38	1.10	0.13	−0.64	−0.13	−2.98	72	0.12
IX	Control – quality	0.21	1.15	0.14	−0.06	0.47	1.52	72	0.18
X	Control – delivery time	1.11	1.42	0.17	0.78	1.44	6.68	72	0.01
XI	Optimization – flexibility	0.10	1.16	0.14	−0.17	0.37	0.71	72	0.07
XII	Optimization – cost reduction	0.52	1.23	0.14	0.23	0.81	3.63	72	0.01
XIII	Optimization – productivity	−0.45	1.09	0.13	−0.71	−0.20	−3.53	72	0.01
XIV	Optimization – quality	0.14	1.08	0.13	−0.12	0.39	1.08	72	0.08
XV	Optimization – delivery time	1.04	1.33	0.16	0.73	1.35	6.70	72	0.01
XVI	Autonomy – flexibility	−0.14	1.56	0.18	−0.50	0.23	−0.75	72	0.21
XVII	Autonomy – cost reduction	0.29	1.71	0.20	−0.11	0.69	1.44	72	0.51
XVIII	Autonomy – productivity	−0.68	1.52	0.18	−1.04	−0.33	−3.86	72	0.01
XIX	Autonomy – quality	−0.10	1.66	0.19	−0.48	0.29	−0.49	72	0.71
XX	Autonomy – delivery time	0.81	1.67	0.20	0.42	1.20	4.13	72	0.01

decision-making and management capabilities has shown a positive relationship in terms of monitoring, control, and autonomy through Sig. value of 0.01, 0.02, and 0.01 > 0.005 value. However, it has shown negative relationship with monitoring with 0.11 < 0.05 level of significance. The integration into IoT in production and management capabilities has shown a positive relationship in terms of monitoring, optimization and autonomy with Sig. value 0.01, 0.01, and 0.01 > 0.005 value. However, the integration and control of production activity showed a negative relationship with the Sig Value of 0.71 > 0.005 value. Achieving IoT implementation goals and the role of management skills outcomes are indicated in this section. The results in terms of management skills and IoT monitoring objectives have shown a positive correlation with productivity improvement with a Sig Value of 0.000 > 0.005 value. However, it has shown poor relationships in terms of flexibility, cost reduction, quality improvement, and delivery time with Sig. Value 0.25, 0.280, 0.202, and 0.41 < 0.005 value. The control factor has shown a positive relationship concerning cost reduction and reduction of delivery time with Sig Value 0.001 and 0.01 < 0.005 value. Third, optimization has shown a positive relationship in terms of cost reduction, improved productivity, and reduced delivery time with Sig. Value 0.001, 0.001, and 0.001 < 0.005 value of significance. However, it has shown a negative correlation with flexibility with values of 0.07 < 0.005 level of significance. The results related to autonomy capacity also showed positive relationships with improving productivity and reducing delivery time with Sig. Value 0.001 and 0.001 > 0.005 value of value. However, it has shown poor relationships in terms of flexibility, cost reduction, and quality improvement with Sig. Value 0.215, 0.51, and 0.71 < 0.005 level of significance.

5 Discussions

The scope of IoT is more than just initiation in production; there were two research areas, i.e., the relationship between IoT use and the purpose of IoT adoption in SMEs, and, secondly, research was related to management skills required for IoT implementation and mapping for IoT implementation purposes. The results regarding IoT use and its relationship to IoT objectives indicate that the production process is closely related to management skills that are monitoring, control, efficiency, and independence, and the same is evident in terms of improved decision-making. Performance in the SME production system. However, a few areas need attention in order to get the full benefit of IoT implementation in SMEs; these are better information flow through the latest RFID sensors, better networks, and equipment integration, and this can improve performance and improve management capacity. SMEs are reaping the benefits of IoT implementation in the organizations. Therefore, the research question about IoT use and management capabilities reflects a moderate level of interaction and requires further improvements in information flow and improved performance. Another research idea was about the administrative capacity and objectives of IoT implementation. The results show that management capacity needs to be improved to achieve the goal of IoT implementation in SMEs. Even if the operating system is moderate, this indicates that SMEs need to develop management skills in order to successfully use IoT. In addition, especially in terms of efficiency and independence, it needs to be understood from the perspective of the successful implementation of IoT in SMEs. Another role of successful IoT adoption in SMEs is related to the identification of the appropriate purpose for IoT implementation. Many research firms were unclear about target selection; in most research SMEs had multiple objectives, so there was a limited level of success in IoT use.

6 Conclusions

To achieve a competitive advantage in the manufacturing process, manufacturing companies need to strive to improve the production process using the latest technology and tools. Current research shows that SMEs have used IoT technology to improve the production process. Research results have shown that SMEs need to define and understand drawing relationships concerning the purpose of IoT use in SMEs. The relationship between IoT use and management capacity indicates that SMEs need to develop management skills to effectively use IoT, although the results indicate that clarity needs to be improved towards the goal of adopting IoT in SMEs. This will support the effective use of IoT. Other research will provide more detailed information on the board environment that improves performance in other SME workplaces, such as foundry, machining, and auto parts. This will provide a deeper understanding of the factors that contribute to IoT use in SMEs.

References

1. M. Ghobakhloo, Determinants of information and digital technology implementation for smart manufacturing. Int. J. Prod. Res. **58**(8), 2384–2405 (2020). https://doi.org/10.1080/00207543.2019.1630775
2. M. Ehret, J. Wirtz, Unlocking value from machines: Business models and the industrial internet of things. J. Mark. Manag. **33**(1–2), 111–130 (2017)
3. H. Yang, S. Kumara, S.T.S. Bukkapatnam, F. Tsung, The internet of things for smart manufacturing: A review. IISE Trans. **51**, 1190–1216 (2019). https://doi.org/10.1080/24725854.2018.1555383
4. I. Akyildiz, J. Jornet, The internet of nano-things. IEEE Wirel. Commun. **17**, 58–63 (2010). https://doi.org/10.1109/MWC.2010.5675779
5. A. Whitmore, A. Agarwal, L. Da Xu, The internet of things—A survey of topics and trends. Inf. Syst. Front. **17**, 261–274 (2015). https://doi.org/10.1007/s10796-014-9489-2
6. A. Bytniewski, K. Matouk, A. Rot, M. Hernes, A. Kozina, Towards industry 4.0: Functional and technological basis for ERP 4.0 systems, in *Studies in computational intelligence*, vol. 887, (Springer, 2020), pp. 3–19
7. M. Chui, M. Löffler, Roberts, R.:The internet of things. Mc Kinsey Q. **2**, 1–9 (2010)

8. R.Y. Zhong, W. Ge, Internet of things enabled manufacturing: A review. Int. J. Agile Syst. Manage. **11**(2), 126–154 (2018)
9. C.L. Constantinescu, E. Francalanza, D. Matarazzo, O. Balkan, Information support and interactive planning in the digital factory: Approach and industry-driven evaluation. Procedia CIRP **25**, 269–275 (2014) Elsevier
10. R. Kumar, R.K. Singh, R. Shankar, Critical success factors for implementation of supply chain management in Indian small and medium enterprises and their impact on performance. IIMB Manag. Rev. **27**, 92–104 (2015). https://doi.org/10.1016/j.iimb.2015.03.001
11. G. MacKerron, M. Kumar, V. Kumar, A. Esain, Supplier replenishment policy using e-Kanban: A framework for successful implementation. Prod. Plan. Control **25**, 161–175 (2014). https://doi.org/10.1080/09537287.2013.782950
12. M. Xia, T. Li, Y. Zhang, C.W. De Silva, Closed-loop design evolution of engineering system using condition monitoring through internet of things and cloud computing. Comput. Netw. **101**, 5–18 (2016). https://doi.org/10.1016/j.comnet.2015.12.016
13. L. Ren, L. Zhang, F. Tao, C. Zhao, X. Chai, X. Zhao, Cloud manufacturing: From concept to practice. Enterp. Inf. Syst. **9**, 186–209 (2015). https://doi.org/10.1080/17517575.2013.839055
14. B. Denkena, B. Dengler, K. Doreth, C. Krull, G. Horton, Interpretation and optimization of material flow via system behavior reconstruction. Prod. Eng. **8**, 659–668 (2014). https://doi.org/10.1007/s11740-014-0545-z
15. Y. Hao, P. Helo, The role of wearable devices in meeting the needs of cloud manufacturing: A case study, in *Proceedings of the Robotics and Computer-Integrated Manufacturing; Elsevier Ltd.*, vol. 45, (2017), pp. 168–179
16. M. Chalal, X. Boucher, G. Marques, Decision support system for servitization of industrial SMEs: A modelling and simulation approach. J. Decis. Syst. **24**, 355–382 (2015). https://doi.org/10.1080/12460125.2015.1074836
17. A. Bonfanti, M. Del Giudice, A. Papa, Italian craft firms between digital manufacturing, open innovation, and Servitization. J. Knowl. Econ. **9**, 136–149 (2018). https://doi.org/10.1007/s13132-015-0325-9
18. D.M. Segura Velandia, N. Kaur, W.G. Whittow, P.P. Conway, A.A. West, Towards industrial internet of things: Crankshaft monitoring, traceability and tracking using RFID. Robot. Comput. Integr. Manuf. **41**, 66–77 (2016). https://doi.org/10.1016/j.rcim.2016.02.004
19. H. Gebauer, E. Fleisch, An investigation of the relationship between behavioral processes, motivation, investments in the service business and service revenue. Ind. Mark. Manag. **36**, 337–348 (2007). https://doi.org/10.1016/j.indmarman.2005.09.005
20. A. Moeuf, R. Pellerin, S. Lamouri, S. Tamayo-Giraldo, R. Barbaray, The industrial management of SMEs in the era of industry 4.0. Int. J. Prod. Res. **56**, 1118–1136 (2018). https://doi.org/10.1080/00207543.2017.1372647
21. A.V. Barenji, R.V. Barenji, D. Roudi, M.A. Hashemipour, Dynamic multi-agent-based scheduling approach for SMEs. Int. J. Adv. Manuf. Technol. **89**, 3123–3137 (2017). https://doi.org/10.1007/s00170-016-9299-4
22. S. Behbahani, C.W. de Silva, System-based and concurrent design of a smart mechatronic system using the concept of mechatronic design quotient (MDQ). IEEE/ASME Trans. Mech. **13**, 14–21 (2008). https://doi.org/10.1109/TMECH.2007.915058
23. S. Brad, M. Murar, Employing smart units and servitization towards reconfigurability of manufacturing processes, in *Proceedings of the Procedia CIRP; Elsevier*, vol. 30, (2015), pp. 498–503
24. J. Bughin, M. Chui, J. Manyika, An executive's guide to the internet of things. McKinsey Q. **4**, 92–101 (2015)
25. P.K. Roberson, S.J. Shema, D.J. Mundfrom, T.M. Holmes, Analysis of paired Likert data: How to evaluate change and preference questions. Fam. Med. **27**, 671–675 (1995)

What Your Tweets Say About You: A Case Study of Extraversion and Word Usage

P. S. Dandannavar, S. R. Mangalwede, and P. M. Kulkarni

Abstract Social networking websites are a popular platform for social interaction and exchange of information and communication. Research in the past has explored the role of the theory of personality and users of social networking sites and successfully established a relationship between users of such sites and personality traits. While most of the work in literature has focused on predicting, with varying degrees of accuracy, the personality traits of Twitter users, the current work focused on determining whether a relationship exists between the personality traits and the nature of tweets. The present study not only explored the application of sentiment analysis and Big Five personality model to examine the personality trait through tweets, but also attempts to establish a relationship between "extraversion" and the polarity of tweets. Results of the study showed that users with extraversion as a dominant trait tend to tweet more positively.

Keywords Personality · Sentiment analysis · Big Five · Personality trait prediction · Personality analysis · Machine learning

1 Introduction

1.1 The Rise of Social Networking Sites

The World Wide Web has now come to become the primary tool that is used by individuals around the globe to create content, connect with one another, and even escape reality. Who remembers life before the Internet? It has become an integral

P. S. Dandannavar (✉) · S. R. Mangalwede
Department of CSE, KLS Gogte Institute of Technology, Belagavi, Karnataka, India
e-mail: padmad@git.edu; mangalwede@git.edu

P. M. Kulkarni
Department of MBA, KLS Gogte Institute of Technology, Belagavi, Karnataka, India
e-mail: pmkulkarni@git.edu

A. Haldorai et al. (eds.), *4th EAI International Conference on Big Data Innovation for Sustainable Cognitive Computing*, EAI/Springer Innovations in Communication and Computing, https://doi.org/10.1007/978-3-031-07654-1_10

and most essential component in our everyday lives. The Internet has influenced almost every aspect of human life – from the way we shop to how we spend our leisure time and even how organizations function [1]. The Internet has had a far-reaching influence in our everyday lives and has integrated so seamlessly into our daily lives that for many of us, it is hard to imagine a life without it. One facet of our everyday lives that has seen major changes due to the Internet has is our social lives [2, 3]. Technological advancements over the years have not only enhanced the nature but also increased the volume of social interaction. One can meet, connect, and interact with co-workers, friends, or strangers from any corner of the world. No other communications technology had made this possible. At the core of this online transformation are social networking sites (SNS), a leading form on online communication. SNS are the most popular sites on the Internet, and their use has grown by leaps and bounds. They are increasingly becoming one of the most popular sources of social networking and sharing of information. SNS allow users to share photos, videos, and other media, to find, discuss, and share news, information, opinions, etc. with people in their network.

As of July 2020, the number of active Internet users stood at a whopping 59% of the global population – that translates to almost 4.57 billion people. Social media is everywhere and has touched almost every aspect of our daily lives. It has also changed our lives in unseen ways – from the way we interact with others to the way we get our daily dose of news. It has become ubiquitous, unavoidable, and powerful, and it is here to stay. Usage of social media has seen exponential growth since 2004. In 2020, an expected 3.6 billion individuals were utilizing online media around the world, a number projected to increase to 4.41 billion in 2025. These numbers are only growing.

1.2 Personality and Social Media

"Social media" – a phrase that is colloquially used these days – refers to "forms of media that allow people to communicate and share information using the Internet or mobile phones" (https://dictionary.cambridge.org/dictionary/english/social-media). Today, people around the globe are connected to each other by virtue of being members of different online social networks [4]. The ubiquitous nature of social media platforms is making them one of the popular and preferred tools for information exchange and social interaction – for individuals, businesses and organizations. There is a large volume of social data (unstructured data) being created from these social activities every second. This social data can be collected and analyzed to derive invaluable insights. For example, marketers can look for customer feedback that may help in increasing sales. Businesses can get a better understanding of how their customers feel about their product or service by analyzing social media posts. This led to the emergence of the field of sentiment analysis (SA), which deals with identification and classification of emotions within textual data.

SA can be applied in diverse fields and hence finds applications. The most prominent SA examples include assessing public sentiment towards products, politics, movies, etc. Users share pictures, music, videos, and articles, as well as their own views, thoughts, and opinions through status updates/comments (FB) and tweets (Twitter). In general, social media platforms have become a medium of expression for users. These expressions are reflections of an individual's personality traits – implying that these user posts can be analyzed to obtain information to recognize personality traits. Personality traits analysis has been an area of active research, and efforts in this direction have proliferated in the past years. Several researchers started to pay attention to this emerging field – applying SA to analyzing posts by a user to predict/recognize his/her personality. In order to express themselves to the world, individuals use social media. These expressions describe the behavior of a user and human behavior reflects his/her personality. A user's social media data can thus be used to predict his/her personality [5].

Personality assessment methods of recent times rely on machine learning and user-generated text (from social media sites) in which personality traits are estimated from plain text by a computer software [6]. Various studies have shown that language use can be considered as a marker of personality because word usage does reflect personality, i.e., personality is revealed in language use [7]. Therefore, it would not be incorrect to say that the character of a person is expressed in their use of language on social networks. Researchers have thus focused on identifying/establishing relationships between different aspects of personality and SNS.

2 Personality and the Internet: A Review

Hundreds of research studies have tried to explore how personality and SNS are related. Previously, it was not thought that language usage showed personality dimensions [8]. Several studies have subsequently tried to explore the relationship between personality and social media behavior. Authors in [9] suggested that Internet use is personality-related. According to [2], in understanding why people behave the way they do on the Internet, personality is a leading factor. Using a number of different theories of personality, this relation between personality and Internet use has been illustrated. Results have revealed that there is a connection between an individual's personality and his/her online behavior and also that personality traits manifest through the language/word usage of users. One of the earliest works on automated detection of personality was published by Argamon et al. [10]. By applying techniques such as Naive Bayes and sequential minimal optimization on casual written text, the authors focused on determining the neuroticism and extraversion dimensions of personality.

The first literature review that examined the relationship between extraversion and social media use was published in 2020 [11]. The authors studied the various ways extraversion is related to social media to summarize that trait characteristics related to extraversion (desire to socially interact, tendency to display positivity) were present in online behavior. While most studies focus on determining personality traits on individual user(s), authors in [6] experimented with celebrity groups. Using tweets from celebrities accounts as input to analyze, they were able to show that the traits of successful celebrities in different professional groups (actors, politicians, writers, businessmen, and sportsmen) differ based on their profession. Descriptive language has long been used to identify different types of personality – English being the predominant. Efforts have been made to establish a correlation between personality traits and other languages. Researchers have been able to successfully demonstrate that personality manifests in linguistic variables. A research was carried out to validate the KLIWC as a personality indicator and to investigate the relationship between language use and personality using a Korean sample [7]. The Korean version of LIWC was used for the analysis to conclude that personality traits were significantly correlated with linguistic variables. A similar work was carried for Filipino Twitter users [12]. Text and personal data was collected from 250 Filipino Twitter users to create personality trait (PT) models based on the way they tweet. The findings of their work indicated that the conscientiousness trait was the easiest to model, followed by extraversion and openness, the hardest trait to model. 115 FB users and 166 nonusers made up the sample of Australian users who were the subjects of a study [13] which aimed to identify the personality characteristics associated with a FB user/nonuser. Results showed that FB users are more extraverted and narcissistic; FB nonusers are more likely to be conscientious, shy, and socially lonely. "Ara personality" – a PT dataset gathered from Egyptian Twitter users – was the contribution of the work [14]. An attempt to predict PT through microblogging behavior was made by [34] for 549 Chinese users. Results did reveal that PT of active users could be predicted by microblogging behavior.

Majority of the research works have focused on establishing correlations between social media/language use and each of the Big Five (OCEAN) traits. However a few studies have tried to investigate the relationship with only a subset of traits with perhaps extraversion and neuroticism the most studied ones. Extraversion, openness to experience, and emotional stability (neuroticism) were the three traits investigated in a study of the US adults [15]. They also examined whether age and gender had any role in the dynamics. Results of their study revealed that extraversion and openness were positively related to social media use and emotional stability negatively. A similar study sought to determine the effect of two PT – extraversion and neuroticism – on social media use in participants aged 18 to 74 (Essays, UK. November 2018). Extraversion seemed to have a marginal effect on an individual's overall use of social media. A pioneering study for social media use in Turkey was conducted by [16]. This empirical study tested the relationship between extraversion, neuroticism and self-esteem, and social media use – with extraversion and neuroticism proving to be more central.

Authors [17] examined the relationship between social media and texting in an attempt to test the shallowing hypothesis. Consistent with one of the hypothesis tested, social media usage was positively correlated with extraversion. Investigating the role of personality and SNS further [1] examined the personality correlates of two largest SNS – FB and Twitter. This was the first study linking Twitter use to personality. Results of this work showed personality differences between those who prefer to use FB/Twitter. An experimental study conducted by [13] examined the variation in user behavior on the Internet and the computer. Results of this study indicated that human behavior, i.e., PTs, are expressed differently in offline and online environments. This is because of a complex system of associations and interactions between behavior and online communication. The study by [4] revealed that personality and use of social networks (by individual users) are interrelated. Work of [16] also aimed at automatically detecting user's personalities from text but was based on the lexical approach. The results however still indicated the way in which an individual interacts with others in SNS is affected by his/her personality. Majority of the researchers used SNS like FB and Twitter as SNS for their work, [18] used predictive models to automatically recognize PT of Yahoo! Answers members, and opined that people high on agreeableness show positive emotions on their profile pictures.

Personality factors – openness to experience and neuroticism – predict blogging. This work carried out by [19] surveyed two samples of undergraduate students – those who blogged and those who did not. Assessment of the personality characteristics of the two groups revealed there were significant differences in the Big Five (particularly neuroticism) between those who blog and those who do not. There can be different uses of social media – for interpersonal interaction, news and/or public affairs, etc. [20] studied the effects of personality over the different social media uses. Investigating the association between personality and types of Twitter users were research studies of [21–23]. The work of [24] focused on the exploring the relationship between personality and word usage. 1,15,00 words collected form 700 bloggers were used to identify correlations with word use for the OCEAN traits and lower-order traits as well [25]. PersonalityViz, an interactive visualization tool to help individuals understand social media-based PT, was presented by [25]. To compute the Big Five personality of a person from one's tweets, the framework used the LIWC text tool and BF personality correlations. Automatic recognition of Big Five traits using users status text on FB was attempted by [20]. The relation between social media activity and users level of OCEAN traits using machine learning algorithms was explored by [26].

Studies have shown that PTs are, in fact, related to SNS use and that personality can be automatically inferred from social media data. "Personality traits are significantly associated with social media use" [25]. "Personality is a powerful explainer of social media user" [20]. Of all the PTs, extraversion and neuroticism are the most studied ones, and it would not be wrong to conclude that the most consistent predictor of active social media use in literature is extraversion.

3 Extraversion and Social Media Usage

Extraversion, one of the personality traits of the BF model, is a more broader character attribute that incorporates various more explicit qualities, for example, sociability (enjoy social interactions), assertiveness (confidence), high activity level, excitement seeking, positivity (positive emotions), and impulsivity. In general, extraverted individuals are more strongly interested in interactions with others – both online and offline. It has been found that personality traits are associated with social media use, and extraversion was the strongest predictor of social media use [27]. Social media sites appeal more to extraverts, and they tend to be drawn to SNS.

Extraversion indicates how outgoing and social a person is. Extraversion is one factor – openness to experience and emotional stability being the others – that has been linked to use of Internet-based social applications. Studies have also indicated that innovative, creative/extraverted individuals are more likely to use social media. Extraverted individuals are more frequent users of social networking sites with their desire for social interactions [16]. Extraversion, as a personality trait, is positively related to social media use. Extraverted individuals use social media frequently so are individuals who are open to experience.

Extraverts are likely to display and attract positivity, away from social media and on it. There is proof that extraversion is associated with usage of specific words on social media [28], and extraverts tend to display positivity by using positive words in social media content. Studies have in fact connected extraversion to the utilization of positive words.

The Big Five model is one of the most popular approaches to personality assessment, with its reliability and validity being supported by many studies. It is developed on the lexical approach. Sir Francis Galton developed the lexical method, which counted the number and types of words used to define a particular type of character [29]. Subsequently, Goldberg [30] developed the lexical hypothesis, defined by two postulates, and suggests that personality characteristics are reflected in the words used to describe other people/things, etc. One can thus look to the language/words used by an individual [31] to know the important personality traits. The present study sought to explore the following hypothesis:

H1: Extraverted individuals are likely to tweet more positively.

4 Methodology

4.1 Background

The current work is a continuation of our previous work titled "Predicting the Primary Dominant Personality Trait of Perceived Leaders by Mapping Linguistic cues From Social Media Data onto the Big Five Model" [32]. Prediction of primary dominant trait was attempted by using Twitter data and the BF model. A randomly selected subset of 26 public figures formed the subset of celebrities whose primary dominant trait we attempted to predict using the latest 3000 tweets for each of them. The celebrities chosen were from various fields (sports, business, literature, politics, and entertainment). Three supervised ML algorithms, viz., Multinomial Naive Bayes, support vector machine, and random forest tree (RFT), were employed. The myPersonality dataset and the NRC emotion lexicon were used to train the models. Reliability of the OCEAN scores predicted by these algorithms was measured using the Cronbach's alpha test. To test the machine learning models, metric measurements, viz., F1-score, Accuracy, Precision, and Recall, were used. The RFT method scored over the other methods both on the Cronbach's alpha test and the metric values and hence its results considered.

Extraversion emerged as the dominant trait of leaders. Conscientiousness, openness to experience, agreeableness, and lastly neuroticism followed, which synched with the work of [33]. Extraversion was, hence, most strongly related with leadership.

4.2 Current Work

For this work, the selected subset of 26 public figures and their tweets used for our previous work was retained. In order to test H1, sentiment analysis (SA) was performed on the tweets of each user using TextBlob.

TextBlob is a simple library in python for NLP on textual data and uses the NLTK library to achieve its tasks. NLTK allows access to several lexical resources and also allows users to work with categorization, classification, and many other tasks easily. TextBlob returns a value for the polarity (between $[-1, 1]$) and subjectivity of a term while evaluating a piece of textual data. The polarity of 1 determines a positive feeling, and a negative feeling is characterized by -1. The sequence of steps involved in detecting the polarity of tweets is shown in Fig. 1.

5 Results and Discussions

For each celebrity considered, "extraversion" had emerged as the primary dominant trait in our previous work. In the context of social media usage, extroverts are described as individuals who are drawn to social media (enjoy social interactions), are frequent users of social media (socializing through online media appeals to extroverts), are energetic and engaged (reflected in their postings of social media material), tend to have a bigger circle/network of friends, and exhibit positivity (using positive emotion words) through using social media. In short, sociometrically popular individuals are extroverts [26].

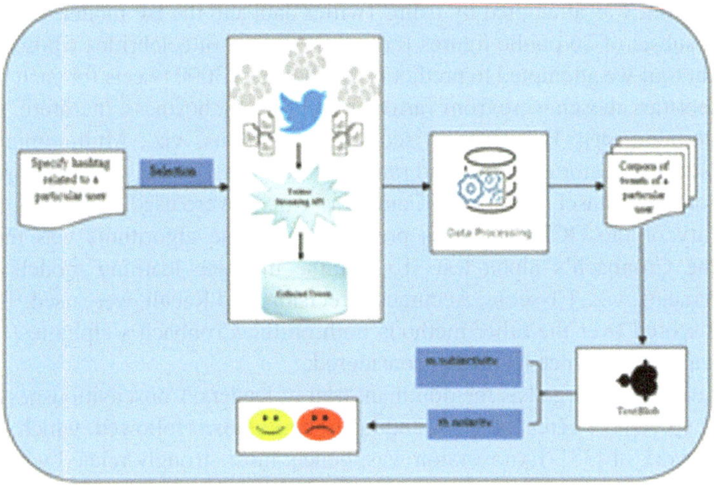

Fig. 1 Process of computing sentiment scores of tweets using the TextBlob

While there has been evidence in literature about extroversion being related with usage of specific word on social media and extroverted individuals displaying positivity, in our opinion, there has been no work that tried to establish an association between personality traits of an individual and the nature of tweets. In this work, we attempted to determine if "extraversion" had an effect on the nature (positive/negative) of tweets, based on the evidence available for "extraversion" in literature and the lexical hypothesis.

The tweets of each user were analyzed, using sentiment analysis, to identify their polarity. Polarity classification assigns a positive or a negative sentiment to an entire tweet. The results of polarity detection for the collection of tweets of each user are depicted in the graph in Fig. 2. The number of positive sentiment tweets outnumbers the number of negative sentiment tweets for each user, which is a clear indication that extraverted individuals tend to tweet more positively. It can thus be concluded that individuals with high scores on extraversion are likely to tweet more positively. Hence H1 can be accepted.

6 Conclusion

One of the personality traits of the BF model is "Extraversion" and includes sub-facets like sociability, assertiveness, high activity level, excitement seeking, positivity and impulsivity. An individual who is extraverted is more outgoing and social – implying that these individuals are interested in interactions with others, whether online or offline. Studies have shown that extraversion is related to social media usage. Extraverted individuals with their desire to socially interact

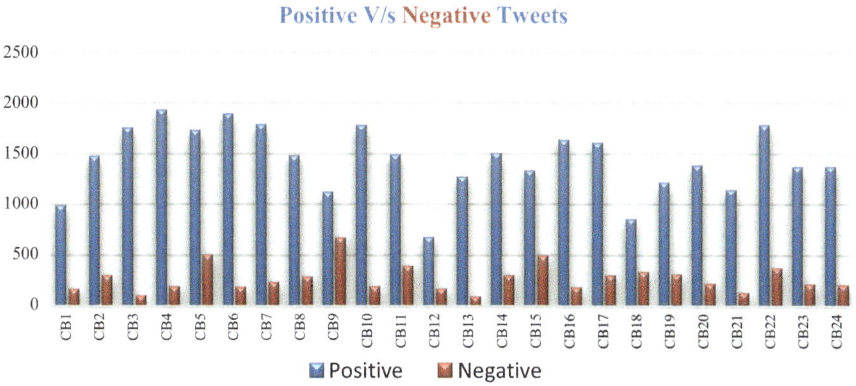

Fig. 2 Graph of polarity of tweets

are drawn to using social media on a regular basis. Extraverts are also likely to display positivity whether on social media or away from it. There is evidence that extraversion is associated with usage of specific words on social media and extraverts tend to use positive words in social media content. Studies have in fact connected extraversion to the utilization of positive words. This work attempted to determine if "extraversion," as a trait, has an effect on the polarity (positive/negative) of tweets. The basis for the hypothesis testing was the lexical hypothesis and the literature available for "extraversion." SA was performed on tweets using TextBlob and NLTK. Results of SA revealed that the number of tweets with positive polarity/sentiment outnumbered the number of tweets with negative polarity/sentiment for all users. This clearly indicates that individuals who are extraverted tend to tweet more positively.

References

1. D.J. Hughes, M. Rowe, M. Batey, A. Lee, A tale of two sites: Twitter vs. Facebook and the personality predictors of social media usage. Comput. Hum. Behav. **28**(2), 561–569 (2012)., ISSN 0747-5632
2. Y. Amichai-Hamburger, G. Wainapel, S. Fox, On the Internet no one knows I'm an introvert': Extroversion, neuroticism, and Internet interaction. CyberPsychol. Behav. **5**, 125–128 (2002)
3. Y. Amichai-Hamburger, E. Ben-Artzi, Loneliness and Internet use. Comput. Hum. Behav. **19**, 71–80 (2003)
4. Y. Amichai-Hamburger, G. Vinitzky, Social network use and personality. Comput. Hum. Behav. **26**(6), 1289–1295 (2010) ISSN 0747-5632
5. P.S. Dandannavar, S.R. Mangalwede, P.M. Kulkarni, Social media text – A source for personality prediction, in *2018 International Conference on Computational Techniques, Electronics and Mechanical Systems (CTEMS)*, (Belgaum, 2018), pp. 62–65
6. K. Dutta, V.K. Singh, P. Chakraborty, et al., Analyzing Big-Five personality traits of Indian celebrities using online social media. Psychol. Stud. **62**, 113–124 (2017)

7. C.H. Lee, K. Kim, Y.S. Seo, C.K. Chung, The relations between personality and language use. J. Gen. Psychol. **134**(4), 405–413 (2007)
8. J.W. Pennebaker, L.A. King, Linguistic styles: Language use as an individual difference. J. Pers. Soc. Psychol. **77**, 1296–1312 (1999)
9. Y.A. Hamburger, E. Ben-Artzi, The relationship between extraversion and neuroticism and the different uses of the Internet. Comput. Hum. Behav. **16**(4), 441–449 (2000)., ISSN 0747-5632
10. S. Argamon, S. Dhawle, M. Koppel, J. Pennebaker, Lexical predictors of personality type. Paper presented at the Joint Annual Meeting of the Interface and the Classification Society of North America, St Louis, MO, 8–12 June 2005
11. H. Green, J. Hinds, A. Joinson, How is extraversion related to social media use? A literature review. Personal. Individ. Differ. **164**, 110040 (2020)., ISSN 0191-8869
12. E.P. Tighe, C.K. Cheng, *Modeling Personality Traits of Filipino Twitter Users* (PEOPLES@NAACL-HTL, 2018)
13. T. Blumer, N. Doering, Are we the same online? The expression of the five factor personality traits on the computer and the Internet. Cyber Psychol. J. Psycho. Res. Cyberspace **6**(3) (2012) article 5
14. H.G. de Zúñiga et al., Personality traits and social media use in 20 countries: How personality relates to frequency of social media use, social media news use, and social media use for social interaction. Cyber Psychol. Behav. Soc. Netw. **20**(9), 540–552 (2017)
15. T. Correa, A.W. Hinsley, H.G. de Zúñiga, Who interacts on the Web?: The intersection of users' personality and social media use. Comput. Hum. Behav. **26**(2), 247–253 (2010)., ISSN 0747-5632
16. J.M. Balmaceda, S. Schiaffino, D. Godoy, How do personality traits affect communication among users in online social networks? Online Inf. Rev. **38**(1), 136–153 (2014)
17. L.E. Annisette, K.D. Lafreniere, Social media, texting, and personality: A test of the shallowing hypothesis. Personality and Individual Differences **115**(2017), 154–158 (2016) ISSN 01918869
18. N. Olivares, L.M. Vivanco, A. Figuero, The big five: Discovering linguistic characteristics that typify distinct personality traits across Yahoo! Answers members. Computación y Sistemas **22**(3), 795–807 (2018)
19. R. Guadagno, B. Okdie, C. Eno, Who blogs? Personality predictors of blogging. Comput. Hum. Behav., 1993–2004 (2013)
20. F. Alam et al., *Personality Traits Recognition on Social Network – Facebook* (ICWSM, 2013)
21. M.S. Salem, S.S. Ismail, M. Aref, Personality traits for Egyptian Twitter users dataset, in *Proceedings of the 2019 8th International Conference on Software and Information Engineering (ICSIE '19)*, (Association for Computing Machinery, New York, 2019), pp. 206–211
22. A. Souri, S. Hosseinpour, A.M. Rahmani, Personality classification based on profiles of social networks' users and the five-factor model of personality. Hum. Cent. Comput. Inf. Sci. **8**, 24 (2018)
23. D. Quercia, M. Kosinski, D. Stillwell, J. Crowcroft, Our Twitter profiles, our selves: Predicting personality with Twitter, in *2011 IEEE Third International Conference on Privacy, Security, Risk and Trust and 2011 IEEE Third International Conference on Social Computing*, (Boston, 2011), pp. 180–185
24. T. Yarkoni, Personality in 100,000 words: A large-scale analysis of personality and word use among bloggers. J. Res. Pers. **44**, 363–373 (2010). https://doi.org/10.1016/j.jrp.2010.04.001
25. L. Gou, J. Mahmud, E. Haber, M. Zhou, PersonalityViz: A visualization tool to analyze people's personality with social media, 45–46 (2013). https://doi.org/10.1145/2451176.2451191
26. M. Stankevich et al., *Analysis of Big Five Personality Traits by Processing of Social Media Users Activity Features* (DAMDID/RCDL, 2018)
27. T. Correa, I. Bachmann, A. Willard Hensley, H. Gil de Zúñiga, Personality and social media use, in *Organizations and Social Networking: Utilizing Social Media to Engage Consumers*, ed. by E. Y. Li, S. Loh, C. Evans, F. Lorenzi, (IGI Global, Hershey, 2013), pp. 41–61

28. "Bloggers' reveal personalities with word choice", https://www.colorado.edu/asmagazine/2010/12/01/bloggers-reveal-personalities-word-choice
29. A. Cargill, The lexical hypothesis & the Big Five model. Int. J. Psych. Res., 2 (2019). https://doi.org/10.33425/2641-4317.1025
30. L.R. Goldberg, An alternative description of personality: The big-five factor structure. J. Pers. Soc. Psychol. **59**, 1216–1229 (1990)
31. E. Roivainen, Frequency of the use of English personality adjectives: Implications for personality theory. J. Res. Pers. **47**(4), 417–420 (2013)., ISSN 0092-6566
32. P.S. Dandannavar, S.R. Mangalwede, P.M. Kulkarni, Predicting the primary dominant personality trait of perceived leaders by mapping linguistic cues from social media data onto the big five model, in *Advanced Machine Learning Technologies and Applications. AMLTA 2020*, Advances in Intelligent Systems and Computing, ed. by A. Hassanien, R. Bhatnagar, A. Darwish, vol. 1141, (Springer, Singapore)
33. A.J. Timothy, J.E. Bono, R. Ilies, M.W. Gerhardt, Personality and leadership: A qualitative and quantitative review. J. Appl. Psychol., Copyright 2002 by the American Psychological Association, Inc. **87**(4), 765–780 (2002) 0021-9010/02/$5.00

Solar Radiation Prediction Using the Random Forest Regression Algorithm

Sandeep Kumar Reddy Thota, C. Mala, Praneetha Chandamuri, and Chandana Nooka

Abstract Solar radiation data is critical for solar energy research and development. It is extremely valuable for design and simulation of solar thermal technologies and solar photovoltaic applications because it offers essential data about the energy that impacts the earth. Because conventional energy sources are rapidly decreasing, sustainable energy sources such as solar, biomass, and wind are growing more prominent. Solar energy is the most generally acknowledged renewable energy source since it is abundant all around the planet. As a result, this project examines the model and methodologies utilized to forecast solar radiation. Random forest algorithm is used in solar radiation prediction; this algorithm combines the benefits of both techniques utilized, resulting in faster calculation and convergence to optimal results.

Keywords Solar radiation · Prediction · Random forest regression · Decision trees · Accuracy · Renewable energy

1 Introduction

Solar power has gained significance over time as a result of the continued natural resources depletion and rising pollution caused by conventional power generation, 63.7 percent of the 350 Gigawatts established capacity of the Indian power system is powered by fossil fuels, whereas 21.2 percent is powered by renewable energy

S. K. R. Thota (✉) · P. Chandamuri · C. Nooka
Department of Computer Science and Engineering, Rajiv Gandhi University of Knowledge Technologies- RK Valley Campus, Kadapa, Andhra Pradesh, India
e-mail: skrthota@gmail.com; praneetha140@gmail.com; chandanar151504@gmail.com

C. Mala
Department of Computer Science and Engineering, National Institute of Technology, Tiruchirappalli, Tamil Nadu, India
e-mail: mala@nitt.edu

A. Haldorai et al. (eds.), *4th EAI International Conference on Big Data Innovation for Sustainable Cognitive Computing*, EAI/Springer Innovations in Communication and Computing, https://doi.org/10.1007/978-3-031-07654-1_11

resources [1]. During the creation of electricity from these fossil fuels, natural gas, oil, and coal, methane, carbon monoxide, sulfur dioxide, carbon dioxide, and other greenhouse gases are emitted, and these gases are exceedingly destructive and detrimental to the environment, contributing to global warming. As a result, more renewable penetration is required to address these issues. By 2022, the Indian government wants to raise solar penetration to 175 GW [2]. The government's recent efforts aim to create a smart grid across the country with built-in solar microgrids. Because the electric power system is only stable when demand and supply are balanced, it is exceedingly intricate. Solar energy generation has the disadvantage of being highly variable and weather reliant. Our article focuses on solar radiation forecasting since solar energy creation is inextricably linked to solar radiation [3, 4]. Solar radiation forecasting is a multidimensional problem in which different weather variables have a stronger correlation with radiation at different periods. This demanded the creation of a model that could evaluate many data samples and estimate and forecast the value of the property.

After cloud vector-based models became popular, Han Seung Jang examined 4 years of previous data using satellite pictures and the support vector machine (SVM) technique to estimate solar radiation [5]. SVM and ANN algorithms struggled to predict solar radiation because of its complex multivariate problem with different meteorological variables showing significant correlations to radiation at distinct intervals. This necessitated the development of a prototype which could consider taking multiple data samples and assess the value of each feature for prediction and forecasting. The decision tree approach, also known as random forest, does this by selecting the appropriate characteristic using the Gini index, which renders the prototype more effective and reliable [6]. For solar radiation forecasting, L. Benali [7] used an artificial neural network, a smart persistence, and a random forest hybrid method in Odeillo, France, and discovered that it produces excellent performance with little divergence between actual and forecasted values compared to other techniques. In Malaysia, Ibrahim et al. [8] developed a hybrid approach that uses RF and a firefly algorithm to estimate hourly global sun radiation. M. Abuella did another investigation in Charlotte, USA, using RF and SVR models for solar radiation forecasting. When compared to typical machine learning models, hybrid algorithms are more effective at predicting solar radiation, according to their research [9]. Sunayana Gupta et al. [1] proposed RF-PSO hybrid model which was used to develop a predictive approach for solar radiation prediction. Many such strategies for forecasting solar-radiation have indeed been established, according to Liexing Huang et al. [10], which include conceptual-parameter models, empirical techniques, AI-powered approaches, and satellite collection information. Throughout the daily or monthly solar radiation forecasting, random forest approaches outperformed.

The following are the key contributions of our work:

- This work is based on the random forest hybrid model; we developed a predictive technique for forecast of solar radiation.

- This algorithm combines the benefits of both techniques utilized, resulting in faster calculation and convergence to optimal results.
- This working model's results are displayed, and a precise forecast of solar radiation is obtained.

The paper's outline is as follows: Section 2 describes about random forest regression. Section 3 explains about random forest regression. Section 4 presents the solar radiation prediction using the random forest regression algorithm. Finally, Section 5 concludes the paper.

2 Random Forest Regression

Random forest regression is a supervised approach made up of a number of decision tree models h(W: Θ_P), P = 1, 2, ..., P; this is a strategy of hybrid regression where Θ_P are randomly distributed random vectors that are independent and identical. If W stands for the input variable and S stands for the output variable, the actual dataset is (W, S). Each decision tree is constructed using two mechanisms: sampling and complete split [11]. The first phase consists of two random sampling procedures that divide the random forest input data separated into row and column sampling. The taking back approach is employed in row sampling, and sampled datasets may contain repetitive samples [12]. The number of sampling samples is u, if the input sample is U (u < U). As a result, not all the observations are used in every tree's input samples during training. We choose v (v < <V) from the V characteristics at random for the column sample. We next create a decision tree utilizing the split of the data sampling approach [13]. Here, Fig. 1 represents the random forest regression architecture [14].

A random forest is a machine-learning solution for finding regression and classification problems. It employs ensemble-learning, a strategy for resolving complex issues by integrating many classifiers. A random forest algorithm is composed of several decision trees. The "forest" generated by the random forest approach is trained using bagging or bootstrapping aggregation [15].

2.1 *Classification Problems*

When running random forests on classification data, you should be aware of the Gini index (p_i), or the formula for calculating the connections between nodes on a decision tree branch:

$$\text{Gini} = 1 - \sum_{i=1}^{n} (p_i)^2 \tag{1}$$

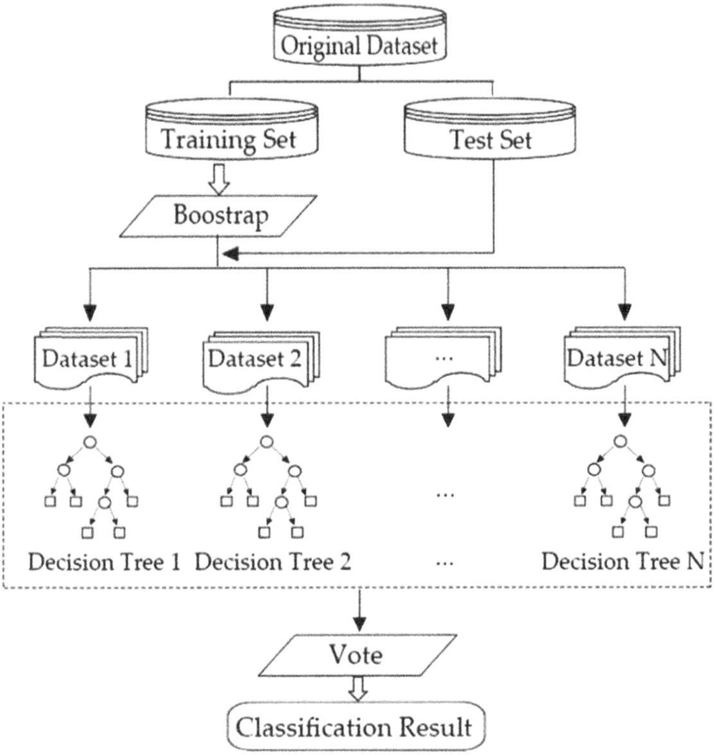

Fig. 1 Random forest regression architecture

2.2 Entropy

Entropy is used to determine how a node should branch based on the likelihood of a specific outcome. It is more mathematically difficult than the Gini index due to the logarithmic function used to calculate it:

$$\text{Entropy} = -\sum_{i=1}^{n} \left(p_i * \log_2 (p_i)\right) \tag{2}$$

2.3 Advantages

- It's a versatile model that can be used for regression and classification challenges.
- Prevents data from being overfit.
- With test data, it's easy to train quickly.

3 Solar Radiation Prediction Using the Random Forest Regression Algorithm

The random forest regression technique is used to forecast solar radiation, and the datasets utilized for prediction are listed below.

3.1 Datasets

These datasets contain meteorological information from the HI-SEAS weather stations collected over a 4-month period (Sep 2016 to Dec 2016) of Mission IV & Mission V [16].

For each dataset, the fields are:

This export's findings can be ordered using a row number ($1 - n$). The UNIX time_t date can be used to sort the findings of this export with those of other exports. The date in yyyy-mm-dd format in the 24-hour format (hh:mm:ss) of the current local time. If there is any numerical data, include it (may be an empty string). If there is any text data, it will be listed here (may be an empty string).

The units of each dataset [15] are depicted in Fig. 2:

All other variables have values at every timestamp throughout each day. There are no other factors that influence the values of time or date. As a result, the date and time of day are unrelated factors. Figure 3 identifies the most significant features in the set.

```
                           Radiation  Temperature  Pressure  Humidity  \
UNIXTime
2016-09-01 00:00:08-10:00       2.58         51.0     30.43       103
2016-09-01 00:05:10-10:00       2.83         51.0     30.43       103
2016-09-01 00:20:06-10:00       2.16         51.0     30.43       103
2016-09-01 00:25:05-10:00       2.21         51.0     30.43       103
2016-09-01 00:30:09-10:00       2.25         51.0     30.43       103

                           WindDirection  WindSpeed  DayLength
UNIXTime
2016-09-01 00:00:08-10:00          77.27      11.25    45060.0
2016-09-01 00:05:10-10:00         153.44       9.00    45060.0
2016-09-01 00:20:06-10:00         142.04       7.87    45060.0
2016-09-01 00:25:05-10:00         144.12      18.00    45060.0
2016-09-01 00:30:09-10:00          67.42      11.25    45060.0
```

Fig. 2 Units of each dataset contain meteorological information

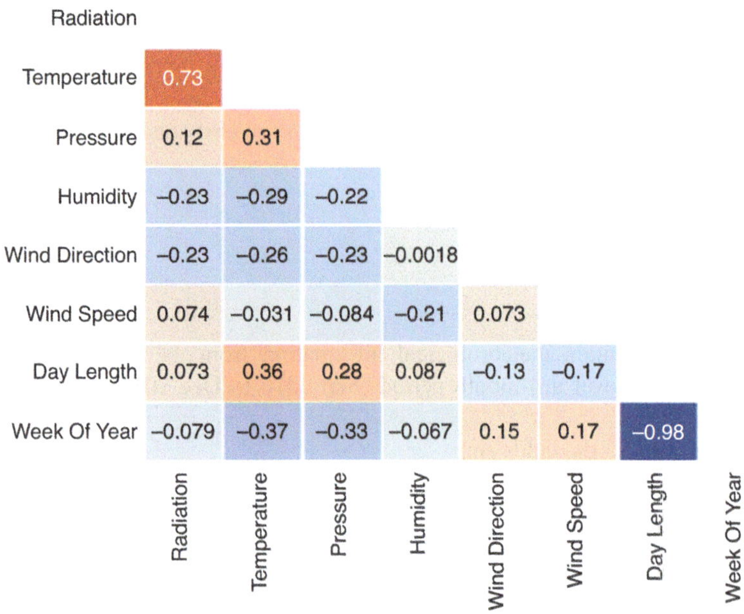

Fig. 3 Identifying the most significant features in the set

```
def plot_test(clf,X_test,y_test):
        y_predicted = clf.predict(X_test)
        p = figure(tools='pan,box_zoom,reset',x_range=[0, 100], title='Mode
        validation',y_axis_label='radiation')
        p.grid.minor_grid_line_color = '#eeeeee'
        p.line(range(len(y_test)),y_test,legend='actual',line_color='blue')
        p.line(range(len(y_test)),y_predicted,legend='prediction',line_color='red')
        output_notebook()
        show(p)
        return
```

Fig. 4 Pseudo code for testing the data

3.2 Implementation

Figure 4 presents the pseudo code for testing the data, Fig. 5. depicts pseudo code for plotting the graph, Fig. 6 indicates pseudo code for training the data, and Fig. 7 depicts pseudo code used for calculating the accuracy.

```
def plot_real(clf,x,y_actual,index):
        y_predicted = clf.predict(x)
        p = figure(toolbar_location='right', title='Predicted vs
        Actual',y_axis_label='radiation',x_axis_type="datetime")
        p.grid.minor_grid_line_color = '#eeeeee'
        p.line(index,y_actual,legend='actual',line_color='blue')
        p.line(index,y_predicted,legend='prediction',line_color='red')
        output_notebook()
        show(p)
        return
```

Fig. 5 Pseudo code for plotting graph

```
def train_model(X,y,clf,debug=False):
        X_train, X_test, y_train, y_test = train_test_split(X,y,test_size=0.2)
        model = clf.fit(X_train,y_train)
        accuracy = clf.score(X_test,y_test)
        return clf, model, accuracy, X_test, y_test
```

Fig. 6 Pseudo code for training data

```
def go(x,y,algorithm,debug=True):
        clf, model, accuracy, X_test, y_test=train_model(x,y,algorithm,debug=True)
        print('Accuracy: %s percent'%str(accuracy*100))

        if debug:
                plot_test(clf,X_test,y_test)
                plot_real(clf,x,y,df.index.values)
                return
```

Fig. 7 Pseudo code for calculating accuracy

3.2.1 Performance Evaluation Metrics

In this study, three statistical variables are utilized to evaluate each prediction scheme: root mean square error (RMSE), mean absolute error (MAE), and accuracy score.

$$\mathrm{RMSE} = \sqrt{\frac{\sum_{i=M+1}^{N} \left(y_i - f\left(x_i\right)\right)^2}{N - M}} \tag{3}$$

$$\mathrm{MAE} = \frac{1}{N - M} \sum_{i=M+1}^{N} \mid \left(y_i - f\left(x_i\right)\right) \mid \tag{4}$$

3.2.2 Testing and Training Data

To avoid bias in the learning process, we use a split train/test methodology to train the algorithm. The dataset is divided into a pool of data points that has been randomly sampled. The first 80% of those points are used for training, whereas the last 20% are used to validate the training data. As a result, the test data isn't always continuous time, but rather a random sample of points from the set.

We use the whole dataset (including training and test points) to illustrate algorithm performance over time for demonstration purposes. This is inherently biased, because some of the points we'll observe have already been trained on and possibly optimized for by the algorithm. However, we check the algorithm accuracy against a subset of testing points (which were not used for training), so we may continue to evaluate performance using the accuracy metric while keeping this potential bias in mind.

3.2.3 Calculating Accuracy

To see how accurate the expected output is, we use accuracy. There are numerous techniques to check the correctness of your linear regression model. Root mean squared error is commonly used. You may train numerous linear regression models, adding or removing features to your dataset, and observe which one has the lowest RMSE – the best one in your scenario. Before fitting the linear regression model, try to standardize your data.

The mean squared error (MSE) is used to illustrate your data branches from each node when utilizing the random forest algorithm to address regression problems:

$$\mathrm{MSE} = \frac{1}{\alpha}\sum_{i=1}^{\alpha} (f_i - y_i)2 \tag{5}$$

Where $\boldsymbol{\alpha}$ is the no. of data points, f_i is the model's output value, and y_i is the data point i's actual value.

The goal of the challenge is to apply Random Forest regression to find the predicted line in the graph and determine the problem's correctness. If the accuracy is too poor, we'll have to update the model.

4 Results and Analysis

4.1 Predicted Verses Actual Graph

Figure 8 represents the output graph for forecasting solar radiation. Here, blue-colored lines indicate actual data representation, and red-colored lines indicate predicted data representation, and Fig. 9 represents the model validation for forecasting solar radiation.

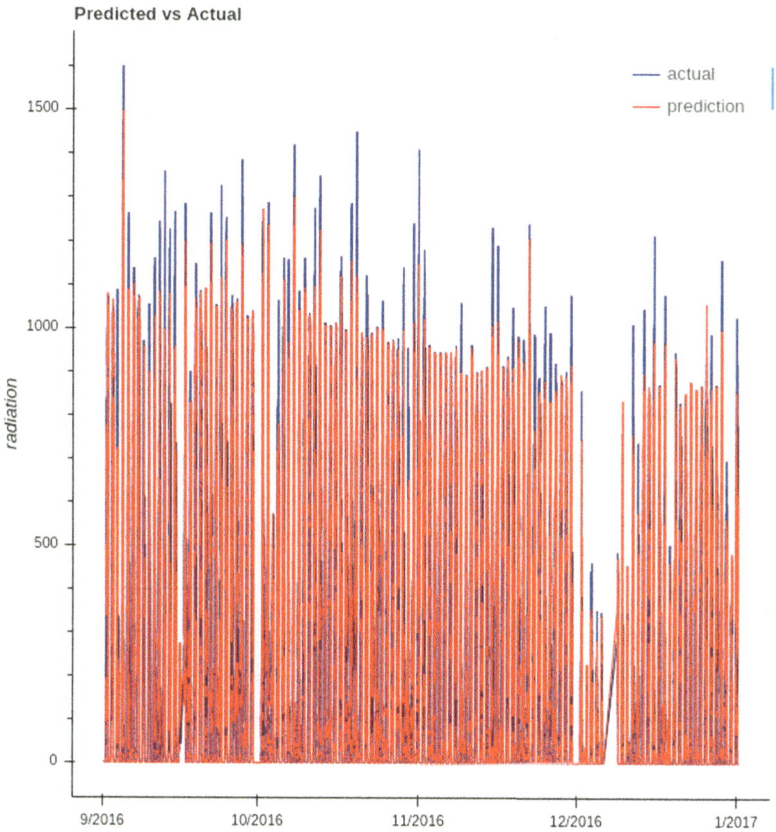

Fig. 8 Output graph for forecasting solar radiation

4.2 *Accuracy*

A machine learning model's accuracy is a measurement that determines whether the system is the most accurate in recognizing patterns and correlations among variables in a dataset based on input or training data. The stronger a method's generalization to 'unobserved' data is, the stronger the prediction that it can generate and the more commercial value it could generate.

The following is the formal definition of accuracy:

$$\text{Accuracy} = \frac{\text{Number of correct predictions}}{\text{Total number of predictions}} \tag{6}$$

Figure 10 represents the accuracy of our model

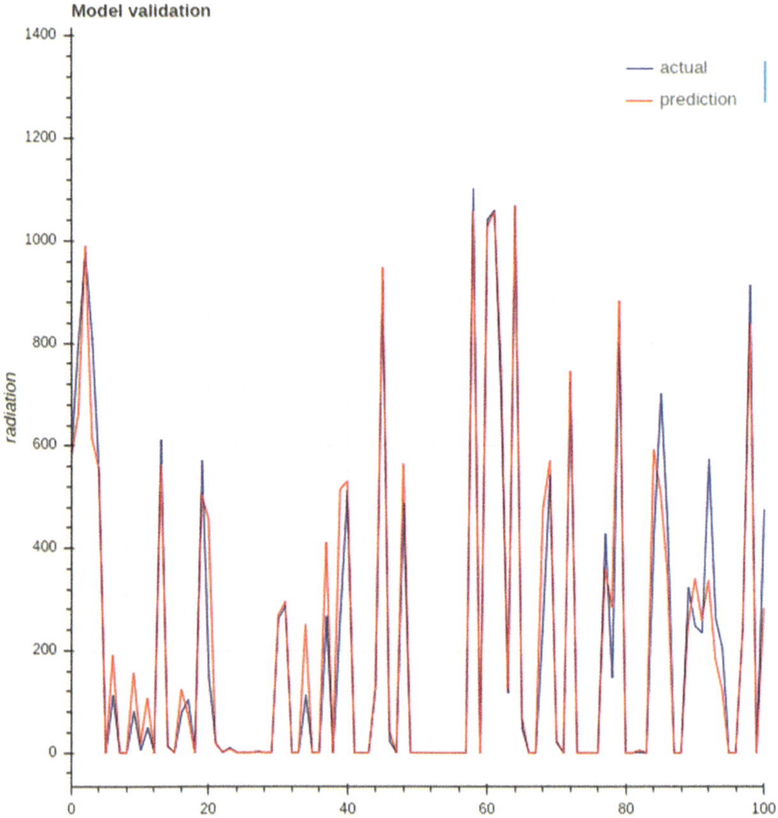

Fig. 9 Model validation for forecasting solar radiation

```
go(x,y,RandomForestRegressor())
Accuracy: 92.7853538859 percent
```

Fig. 10 Accuracy of our model

5 Conclusion

In this work, we developed a predictive technique for solar radiation prediction based on the random forest hybrid model. This approach investigated an optimization technique in conjunction with a robust tree-based model for solar radiation. This algorithm combines the benefits of both the techniques utilized, resulting in faster calculation and convergence to optimal results. The results of this working model are displayed, and a precise prediction of solar radiation is obtained. This technique is effective for developers because it prevents the difficulty of dataset overfitting.

Finally, it is a valuable tool for developing reliable forecasts in strategic decision-making for enterprises.

References

1. S. Gupta, A.R. Katta, Y. Baldaniya, R. Kumar, Hybrid Random Forest and Particle Swarm Optimization Algorithm for Solar Radiation Prediction, in *2020 IEEE 5th International Conference on Computing Communication and Automation (ICCCA)*, (Galgotias University, Greater Noida, 30–31 Oct, 2020)
2. https://niti.gov.in/writereaddata/files/writereaddata/files/
3. D. Chandolaa, H. Guptab, V.A. Tikkiwala, M.K. Bohra, Multi-step ahead forecasting of global solar radiation for arid zones using deep learning, in *International Conference on Computational Intelligence and Data Science (ICCIDS 2019)*, Procedia Computer Science, vol. 167, (2020), pp. 626–635
4. E. Lazarevska, J. Trpovski, A neuro-fuzzy model of the solar diffuse radiation with relevance vector machine, in *11th International Conference on Electrical Power Quality and Utilisation*, (2011)
5. H. S. P. H. S, K.Y. Bae, D.K. Sung, Solar power prediction based on satellite images and support vector machine. IEEE Trans. Sustain. Energy **7**(3), 1255–1263 (2016)
6. L. Random forests, mach learn. (2001).
7. A.C.R.L. Benali, G. Notton, Solar radiation forecasting using artificial neural network and random forest methods: Application to normal beam, horizontal diffuse and global components. Renew. Energy **132**, 871–884 (2019)
8. I.A. Ibrahim, T. Khatib, A novel hybrid model for hourly global solar radiation prediction using random forest technique and firefly algorithm. **138**, 413–425 (2017)
9. B.C. Mohamed Abuella, Random forest ensemble of support vector regression models for solar power forecasting, in *Innovative Smart Grid Technologies, North America Conference*, (2017)
10. L. Huang, J. Kang, M. Wan, L. Fang, C. Zhang, Z. Zeng, Solar radiation prediction using different machine learning algorithms and implications for extreme climate events. Front. Earth Sci. **9** (2021). https://doi.org/10.3389/feart.2021.596860
11. F.E. Yeboah, R. Pyle, Predicting solar radiation for renewable energy technologies – A random forest approach. ResearchGate article. Int. J. Mod. Eng. **16**(1, Fall/Winter) (2015)
12. J. Liu, M.Y. Cao, D. Bai, R. Zhang, Solar radiation prediction based on random forest of feature-extraction. IOP Conf. Ser.: Mater. Sci. Eng **658**, 012006 (2019)
13. S. Mohanty, P.K. Patra, S.S. Sahoo, Prediction and application of solar radiation with soft computing over traditional and conventional approach – A comprehensive review. Renew. Sustain. Energy Rev. (2016)
14. J. Chen, J. Yu, J. Wen, C. Zhang, Z.E. Yin, J. Wu, S. Yao, Pre-evacuation time estimation based emergency evacuation simulation in urban residential communities. Int. J. Environ. Res. Public Health **16**, 4599 (2019). https://doi.org/10.3390/ijerph16234599
15. G. De'ath, Boosted trees for ecological modeling and prediction. Ecology **88**(1), 243–251 (2007)
16. https://www.kaggle.com/runphilrun/hi-seas-solar-radiation-prediction/data

Part IV
Bigdata and Technology

Online Covid-19 Risk Analysis System for Early Detection of Possible Infection

Dahlia Sam, A. Kameshwaran, S. ArunKumar, M. Charumathi, and K. N. Mithilesh

Abstract Coronavirus disease (COVID-19) is an infectious disease caused by the SARS-CoV-2 virus that has been spreading worldwide. Most people infected with the virus experience severe impact in their health, finances, and many other aspects. Infected people often experience mild to moderate respiratory illness and recover without requiring special treatment. However, some become seriously ill and require timely medical intervention. Older people and those with underlying medical conditions like cardiovascular disease, diabetes, chronic respiratory disease, or cancer are more likely to develop serious illness. Anyone can get sick with COVID-19 and become seriously ill or die at any age. As the number of cases continues to rise, it is clear that these viruses pose a threat to public health. In this work, an analysis is performed on whether an individual has been affected with coronavirus or not. This is done via an online portal, where the user answers simple questions regarding the symptoms of the disease like cold, dry cough, loss of taste and smell, tiredness, etc. Once the given questions are answered, the user will be intimated whether he has been possibly infected with coronavirus or if the symptoms are not alarming. In case of possible infections, the application also shows the possible level of infection so that the user can take necessary precautions. The user could also take subsequent measures like visiting the nearby hospital immediately or having a remote consultation with a physician. This concept is made in form of website, where anyone can easily answer the questions and check the results. This method is quite feasible and an effective way of preliminary analysis that can done anywhere and anytime.

Keywords Covid-19 risk level analyzing · Identifying by symptoms

D. Sam (✉) · A. Kameshwaran · S. ArunKumar · M. Charumathi · K. N. Mithilesh
Department of Information Technology, Dr. M.G.R. Educational and Research Institute, Chennai, Tamil Nadu, India
e-mail: dahliasam@drmgrdu.ac.in; kameshwaran.it@drmgrdu.ac.in

A. Haldorai et al. (eds.), *4th EAI International Conference on Big Data Innovation for Sustainable Cognitive Computing*, EAI/Springer Innovations in Communication and Computing, https://doi.org/10.1007/978-3-031-07654-1_12

1 Introduction

In the recent past, the entire world has been affected by the COVID-19 pandemic. However, the impact of the pandemic and its consequences are felt differently depending on our status as individuals and as members of society. While some try to adapt working online, homeschooling their children, and ordering food via Instacart or other online food delivery services, others have no choice but to be exposed to the virus while keeping the society functioning. Our different social identities and the social groups we belong to determine our inclusion within society and, by extension, our vulnerability to epidemics.

COVID-19 has been killing people on a large scale. As of October 10, 2020, more than 7.7 million people across every state in the United States and its four territories had tested positive for COVID-19 [1, 2]. According to *The New York Times* database, at least 213,876 people with the virus have died in the United States [9, 12]. However, these alarming numbers give us only half of the picture; a closer look at data by different social identities (such as class, gender, age, race, and medical history) shows that minorities have been disproportionally affected by the pandemic [3, 5]. These minorities in the United States are not having their right to health fulfilled [6–8].

In order to reduce the impact of the virus, early detection of the viral infection is crucial [4]. This can assist the users to take necessary precautionary measures and also to check the level of possible infection [10, 11]. To aid this, a website is created that can be made available for everyone to access and check the results by answering few simple queries.

2 Scope of the Project

The concept of analyzing whether any individual user has been affected by corona virus or not is done by answering a questionnaire through the website. This contains questions regarding symptoms of the disease like cold, dry cough, loss of taste or smell, tiredness etc. The purpose of this work is to examine whether the user has been affected to coronavirus or not by answering the questions, and the user can check their results according to their answers.

3 Proposed Methodology

The methodology used here is that of analyzing whether the user has been affected to coronavirus or not by answering few simple questions regarding the symptoms of the disease like cold, dry cough, loss of taste and smell, tiredness, etc. Once the given questions are answered, the user will be intimated whether he has been infected with coronavirus or a normal fever, so that they would not panic. It also shows the level of infection so that we can take corresponding measures. This concept is made in form of website, where anyone can answer the questions and check the result.

4 System Architecture

The admin sets the questions related to COVID-19 symptoms, so that the user can view the questions and respond to those questions. Based on the reply given by the user, the risk level is calculated. After that the user will be intimated whether he has been possibly infected with coronavirus or not. Depending on the symptoms entered by the user, the risk level is also analyzed. The guidelines to be followed for different risk levels are also included. These guidelines include all precautionary measures that can be taken at home immediately. After viewing the result, the user can check to their nearest hospitals and can also check the availability of beds in the hospitals. He can also decide to be on quarantine immediately preventing further spread of infection. This architecture is shown diagrammatically in Fig. 1.

5 Module Description

This COVID-19 Risk Analysis system has several sub-modules, each of which has its own functionality. These different modules are listed below.

5.1 ADMIN

- Sets the questionnaire related to COVID-19 symptoms.
- Display information about vaccination and videos to the user.
- Display precautions and guidelines to user.
- Display nearest hospital and availability of bed.

5.2 USER

- Enter name and age.
- Respond to the questions.
- Get the risk level.
- View precautions and guidelines.
- View nearest hospital and availability of bed.

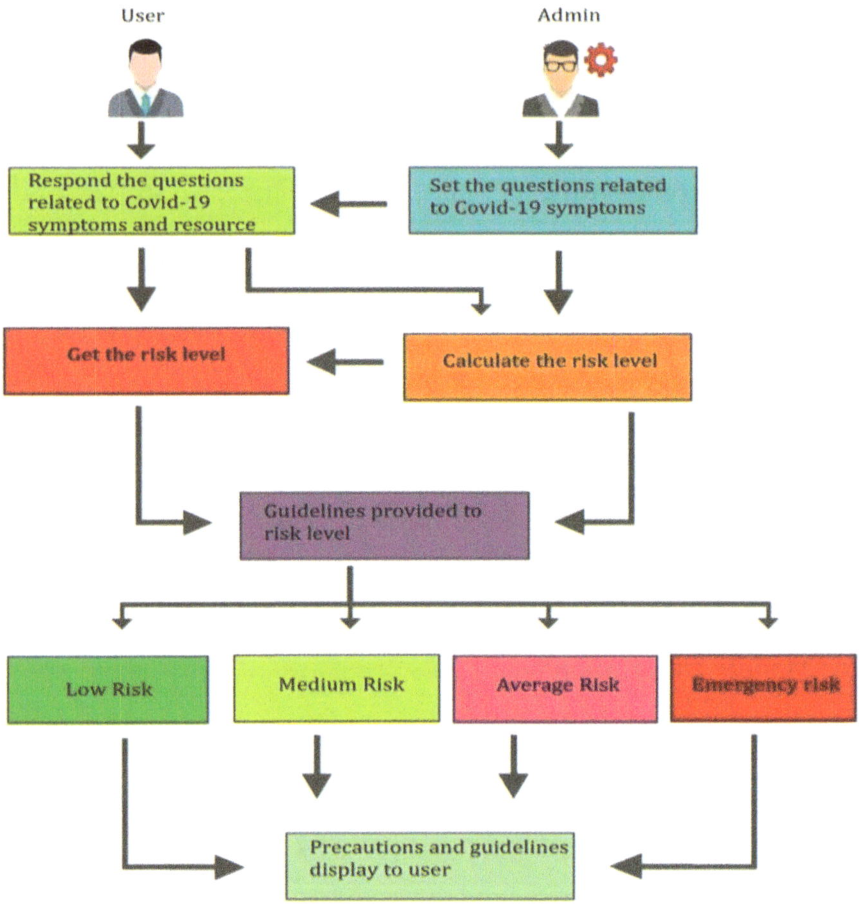

Fig. 1 Architecture diagram

6 Implementation Details

6.1 ADMIN

The admin side will set the questions and provide the precautions to the user.

6.2 USER

The user can respond to the questions and view the result and also check the nearby hospitals.

7 COVID-19 Risk Analysis Questions

- Enter Your Name:
- Enter Your Age:
- Select your Gender:
- Are you a:
- What's your current body temperature?
- What is your saturation rate in oximeter?
- Do you have fever?
- Do you have any throat pain, cold, or cough?
- How long do you have those symptoms?
- How many members are staying with you?
- Are any of your family members or friends facing the same issue?
- Did you get affected by COVID-19 before?
- Has any of your family members or friends been affected by COVID-19 before?
- Has any of your family members or friends been vaccinated for COVID-19?
- Has any of your neighbors been affected by COVID-19?
- What about your daily working routine?
- Do you carry your mask and sanitizer every day with you?
- Do your friends and family members carry their mask and sanitizer every day with them?
- What type of mask are you using?
- Are you a traveler?
- Did you take any tablets for your fever, cold, and cough?
- Did you face any difficulties in breathing or shortness of breathing?
- Do you have any chest pain or pressure?
- Are you a sugar patient?
- Do you keep your body hydrated every time?
- Do you have tiredness?
- Do you invite any of your friends, relatives, or neighbors every day?
- Do you follow social distancing?
- Do you have any sinus problems?
- Do you have any dust allergies?
- Do you have any pets?
- Have you heard about anybody being affected by COVID-19 around 500 m from your living area?
- If yes how many?
- Do you clean your living space every day?
- Do you clean your hands frequently?
- Do you have a separate room?
- Do you share your room less than 1 m apart?
- Have you gone for walking or to the gym for exercise?

8 Results

The proposed system was implemented and tested. The visible information for the users and the emergency helpline numbers are shown in Fig. 2.

The self-declaration form that the user needs to fill out in order to analyze his possible infection level is shown in Fig. 3. The reply given by the user is used for analysis (Figs. 4, 5 and 6).

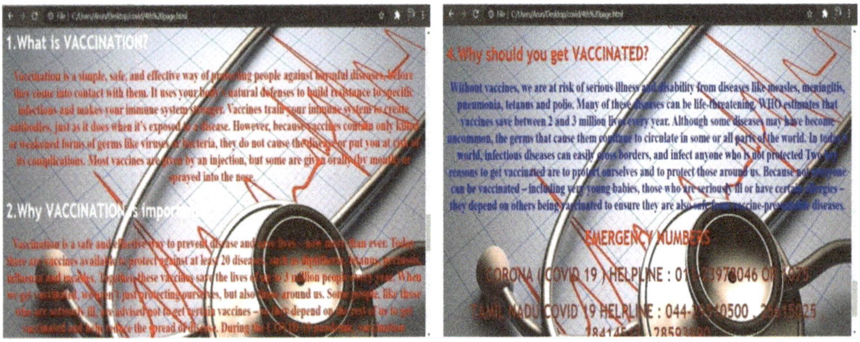

Fig. 2 Importance of vaccination

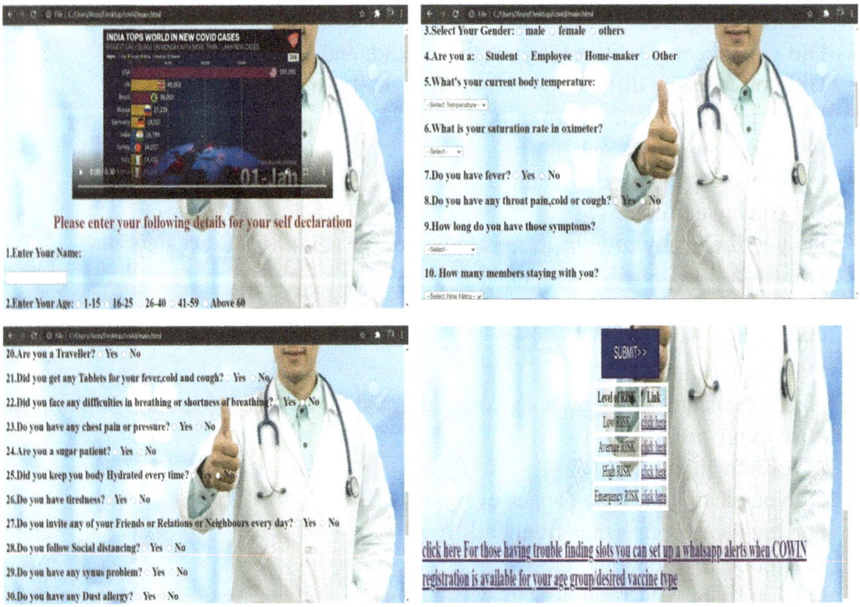

Fig. 3 Self-declaration form

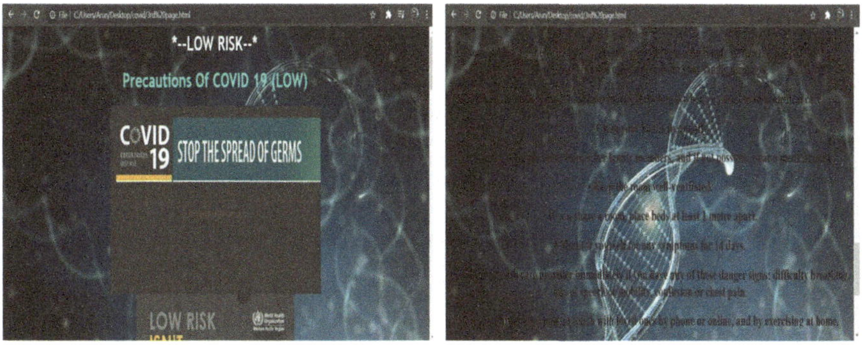

Fig. 4 Low risk

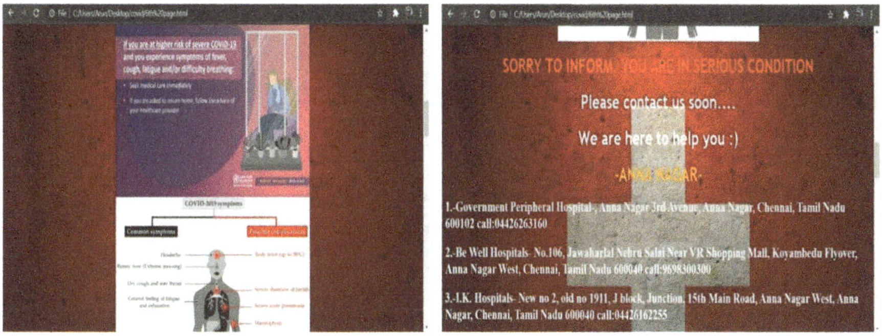

Fig. 5 Emergency risk

9 Results and Discussion

The main concept of the website is to check and analyze whether the user is infected with the virus or not. The web page consists of 30 questions which ask for several symptoms faced by the individual. The individual is asked to enter his/her name and age and followed by the symptoms faced. The user is asked to answer each and every question. After analyzing the data, the user gets the result whether the infection is low, average, high or is at the emergency situation.

9.1 Workshop

In workshop, 30 questions were given, and n numbers of dataset were collected from different people. Individual responses to those questions were entered by the users before submitting it. After that an average was found as shown in Fig. 7. Based on

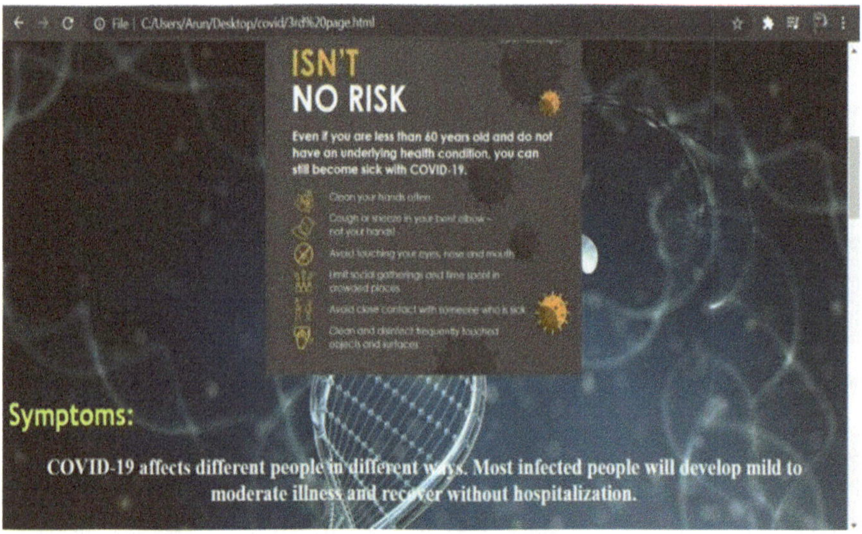

Fig. 6 No risk

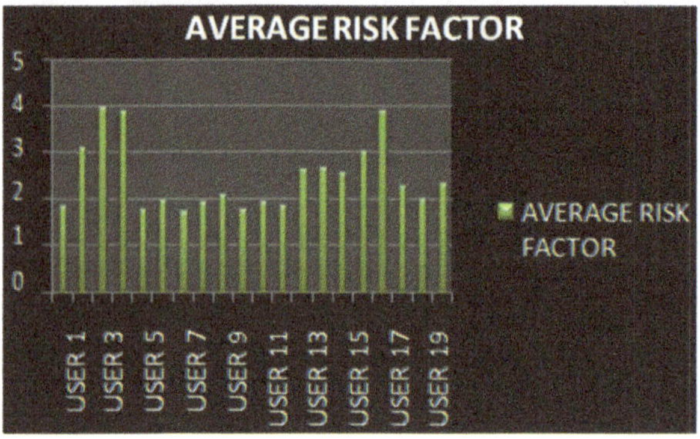

Fig. 7 Average risk factor

this the users risk level can be analyzed. Table 1 shows the workshop-related data and average risk factor obtained.

10 Conclusion

In this work, a webpage is developed for any users to enter his health details. The health condition can be analyzed by data input given by the user. After analysis the user receives the result on whether the infection is low, average, or high or is an emergency. This aids him to take the necessary precautionary steps and also contact

Table 1 The workshop-related data and average risk factor obtained

Users	Average risk factor	Level
USER 1	1.86	Low
USER 2	3.14	High
USER 3	4.00	Emergency
USER 4	3.91	Emergency
USER 5	1.77	Low
USER 6	2.00	Average
USER 7	1.74	Low
USER 8	1.94	Low
USER 9	2.09	Average
USER 10	1.77	Low
USER 11	1.94	Low
USER 12	1.86	Low
USER 13	2.66	Average
USER 14	2.69	Average
USER 15	2.57	Average
USER 16	3.06	High
USER 17	3.94	Emergency
USER 18	2.29	Average
USER 19	2.00	Average
USER N	2.34	Average

the nearby hospital immediately before the viral infection worsens. The COVID-19 pandemic is spreading across the globe at an alarming rate. Timely identification of the infection is the key to getting timely medical help, thereby preventing severity of the disease. This will reduce the possible deaths and also financial loss for the patients. This system will be a boon for all. In the near future, COVID-19 vaccine booking system can be included in the website that will help the people to book their vaccines easily. Another possible future work is adding pharmacy system in the website that will help the people to purchase medicines from the place where they live, so that they do not have to come outside during the pandemic. This system is surely a blessing to all the people considering the current scenario.

References

1. S. Hassan, F.N. Sheikh, S. Jamal, J.K. Ezeh, A. Akhtar, Coronavirus (covid-19): A review of clinical features, diagnosis, and treatment. Cureus **12**, e7355 (2020)
2. T. Singhal, A review of Coronavirus Disease-2019(covid-19). Indian J. Pediatr. **87**, 281–286 (2020)
3. K.J. Clerkin, J.A. Fried, J. Raikhelkar, G. Sayer, J.M. Griffin, et al., COVID-19 and cardiovascular disease. Circulation **141**, 1648–1655 (2020)
4. H. Harapan, N. IToh, A. Yufika, W. Winardi, S. Keam, et al., Coronavirus disease 2019(covid-19): A Literature review. J. Infect. Pub. Health **13**, 667–673 (2020)

5. R. Kumar, A. Agarwal, Covid-19: Common prevention-brief review. SSRN Elect. J. (2020)
6. W. Wang, M. Enilov, The global impact of Covid-19 on financial markets. SSNR Elect. J., 10 (2020)
7. D. Patel, V. Patel, Epidemiology, prevention, and management of covid-19. Crit. Rev. Phys. Rehab. Med. (2020)
8. A.A. Milibari, Current situation of coronavirus disease: (COVID-19) review article. Health Sci. J. Sp. Iss **1**, 005 (2020)
9. Y. Shoenfeld, Corona (COVID-19) time musings: Our involvement in COVID-19 pathogenesis, diagnosis, treatment and vaccine planning. Autoimmun Rev **19**, 102538 (2020)
10. F. Di Gennaro, D. Pizzol, C. Marotta, M. Antunes, V. Racalbuto, et al., Coronavirus Diseases (COVID-19) current status and future perspectives: A narrative review. Int. J. Environ. Res. Pub. Health **17**, 2690 (2020)
11. E. Padron-Regalado, Vaccines for SARS-CoV-2: Lessons from other coronavirus strains. Infect. Dis. Ther. **9**, 255–274 (2020)
12. F. He, Y. Deng, W. Li, Coronavirus disease 2019 (COVID-19): What we know? J. Med. Virol. **92**, 1–12 (2020)

Finger Knuckle Print Recognition Using Complex Conjugate Feature Vector

P. Jayapriya and K. Umamaheswari

Abstract Biometrics are human-specific traits that are employed in person identification and access control. Various biometrics such as fingerprints, finger knuckle, iris, palmprint, vein patterns, and DNA are used in recognition. Among these numerous biometrics, the researcher is drawn to the hand-based finger knuckle print (FKP). On the dorsal side of the hand, the FKP biometric is found. The creases and folds in the finger knuckle print are rich in textural pattern and can be utilized to identify individuals. The recognition of finger knuckles based on a complex feature fusion is proposed in this chapter. The subspace techniques such as principle component analysis (PCA) and linear discriminant analysis (LDA) are used to extract complex number features. These extracted complex vectors are fused using a parallel fusion strategy. Finally, finger knuckle PolyU and IIT Delhi datasets are used to test the developed parallel fusion complex features. The experimental results show that the proposed parallel fusion of complex vector for feature extraction techniques improve the classification accuracy.

Keywords Feature extraction · PCA · LDA · Recognition · Finger knuckle print

1 Introduction

Biometric characteristics are used in different fields to authenticate the person. To identify an individual, biometric traits such as physical or behavioral traits are used [1]. The physiological characteristics are the most static, not very variable in time, while the behavioral characteristics are subject to variations. It may also be influenced by external factors or by specific emotional conditions such as stress or high psychological repercussions. Researchers are interested in hand-based biometric characteristics such as FKP, fingerprint, palmprint, finger vein, palm vein, and hand geometry because they are accurate and concise to implement with low

P. Jayapriya (✉) · K. Umamaheswari
Department of Information Technology, PSG College of Technology, Coimbatore, India

A. Haldorai et al. (eds.), *4th EAI International Conference on Big Data Innovation for Sustainable Cognitive Computing*, EAI/Springer Innovations in Communication and Computing, https://doi.org/10.1007/978-3-031-07654-1_13

cost [2, 3]. The biometric finger knuckle print (FKP) is employed for identification and verification in this work. Person identification is based on the skin shape of the outermost layer of the finger around the phalangeal joint. It has invariant, quantifiable, adequate, and stable properties. It is small in size when compared with other hand-based biometrics [4].

The features that scored high in face biometric recognition were extracted using linear projection algorithms such as PCA and LDA [5]. PCA is a subspace algorithm which preserves the class discriminate information as much as possible. LDA is a linear combination of features used to decrease the amount of features while relying on class separability. These two appearance-based algorithms have their own advantage and disadvantage. PCA is used for select feature that represents the class and LDA for class separability. PCA works well for a limited number of samples per class, whereas LDA performs better for a large number of samples. To reap the benefits of both algorithms, the features are extracted using PCA and LDA [6]. The apparent strong correlation between LDA and PCA when frontal views are used and PCA is applied before LDA denied the fusion of such algorithms. Whenever the LDA transformations are performed to the principal components, it can result in a feature space that is substantially different from the PCA feature space; so take this into consideration when comparing LDA with PCA [7]. Theoretical and experimental research into the integration of LDA and PCA for finger knuckle biometric identification and verification is encouraged as a result. Fusion can be classified into three types: centralized data fusion, decision-level fusion, and feature-level fusion [8].

The proposed work uses the feature fusion. It is the fusion of multiple sets of features for classification. The advantages of using the feature-level fusion are that it originates the discriminant features from the multiple sets of features and also eliminates the redundant features to form the correlation among the distinct features. The classical method for feature fusion is divided into two parts: serial fusion, which is simply the union of two feature sets, and parallel fusion, which is the union based on complex vectors [9].

This chapter focuses on the integration of PCA and LDA for finger knuckle pattern classification. In reality, classification is correlated to the amount of classes and feature sets, with recognition improving as the number of features grows. Parallel feature fusion depending on a complex feature vector is proposed for finger knuckle recognition. The experimental results for parallel feature fusion based on complex feature vector reveal that classification accuracy improves as the number of features decreases. The experimental results for the proposed feature fusion show that classification accuracy improves as the number of features reduces.

The rest of this article is structured as follows. Section 2 deliberates literature study. Section 3 presents the proposed finger knuckle recognition founded on feature extraction and fusion techniques. Section 4 presents the outcomes of the experiments and discusses the detailed experimental comparisons. Section 5 concludes with a conclusion.

2 Literature Study

Woodard and Flynn [10], for example, have effectively demonstrated the benefit of using 3D finger dorsal images for personal identification. The authors have created a curvature-based identification method that uses a 3D sensor to capture a 3D finger surface.

Kumar et al. [11] developed an online system that combines finger knuckle patterns from numerous fingers and geometrical shape parameters using subspace-based approaches such as principal component analysis (PCA), independent component analysis (ICA), and linear discriminant analysis (LDA). In [12, 13], the Gabor filter is employed to retrieve finger knuckle features. To reduce dimension, kernel principal component analysis and linear discriminant analysis are applied.

Chlaoua et al. [14] newly explored the use of a principal component analysis network (PCANet) in finger knuckle print feature extraction. The PCANet is constructed on a direct deep learning topology. PCANet was used to examine two phases of filter banks, which were charted using a simple binary hashing and block histograms enabling feature vector clustering. Finally, for categorization, a linear multiclass support vector machine (SVM) was used. Furthermore, Chalabi et al. [15] developed a system based on score-level fusion of minor and major finger knuckles using the PCANet-SVM approach.

Although principle component analysis (PCA) is a popular and simple method for dimensionality reduction, it ignores class separation [16]. Linear discriminative analysis (LDA) [17] can be utilized to circumvent PCA's issues and enhance feature subspace separability, resulting in appealing recognition stage performances. LDA, on the other hand, continues to presume that all classes have the same covariance matrix, which contradicts the normality principle. In this research, fusion of PCA and LDA was utilized to overcome some of the limitations of PCA and LDA while maximizing their strengths, with the PCA method used to minimize the dimensionality of large features and the LDA technique applied on PCA weights to boost the separability among classes.

The combined feature always results in high-dimensional vector which contains irrelevant and redundant information. This results in low recognition rate. Therefore, the feature fusion is done for the combined features. However the fusion is preprocessing process included in feature combination. There are two types of fusion such as parallel fusion and serial fusion. The serial concatenated feature combination, on the other hand, breaks the feature dimension, increasing the classifier's learning and recognition times significantly, which is unacceptable in tasks needing real time [18] or high efficiency [19]. Scholars have used dimensionality reduction techniques [20–23] to perform different dimensionality reduction and weighting on multiple features to acquire reduced features [24], but this may cause key information to be overlooked, affecting classification accuracy [25]. To solve this issue, the study proposes the parallel fusion of complex vectors based on Euclidean distance classification.

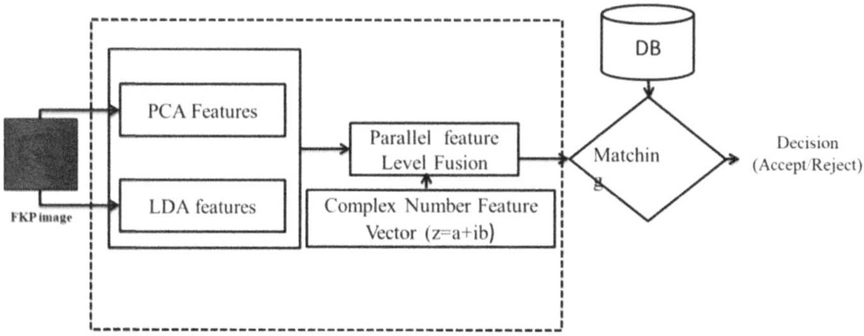

Fig. 1 Proposed FKP recognition system

3 Proposed PCA_LDA Complex Conjugate Vector FKP System

The proposed feature-level fusion based on complex feature vector for FKP recognition is presented in this section. The key points and procedure of parallel and serial fusion are also discussed in this work. The proposed FKP biometric recognition system is shown in Fig. 1. The steps in this procedure were as follows: (i) two subspace algorithms, PCA and LDA, are used to extract the features from the FKP image, (ii) these features are fused at the feature level using a parallel fusion strategy, and (iii) finally k-nearest neighbor (KNN) classifier is used to evaluate the performance metrics of the FKP. The theoretical framework of the FKP proposed work is briefly detailed in the following subsection.

3.1 *Feature Extraction*

Feature extraction plays a significant role in image processing. The features are extracted from the preprocessed image which contains the information of the related image pattern. The efficient feature improves the classification accuracy and recognition rate.

Principle Component Analysis

The aim of using PCA is to transform the higher-dimensional space to lower-dimensional space. Let data be ($X = \{x_1, x_2, \ldots, x_N\}$) transformed from higher-dimensional space (R^M) to lower-dimensional space (R^K), where N is a total number of finger knuckle images, x_i, representing ith sample of image. All samples should have the same dimension which represented as ($x_i \in R^M$). The direction of the PCA space represents the direction of the maximum variance of the given data. The PCA space, as depicted in the figure, is composed of a number of components.

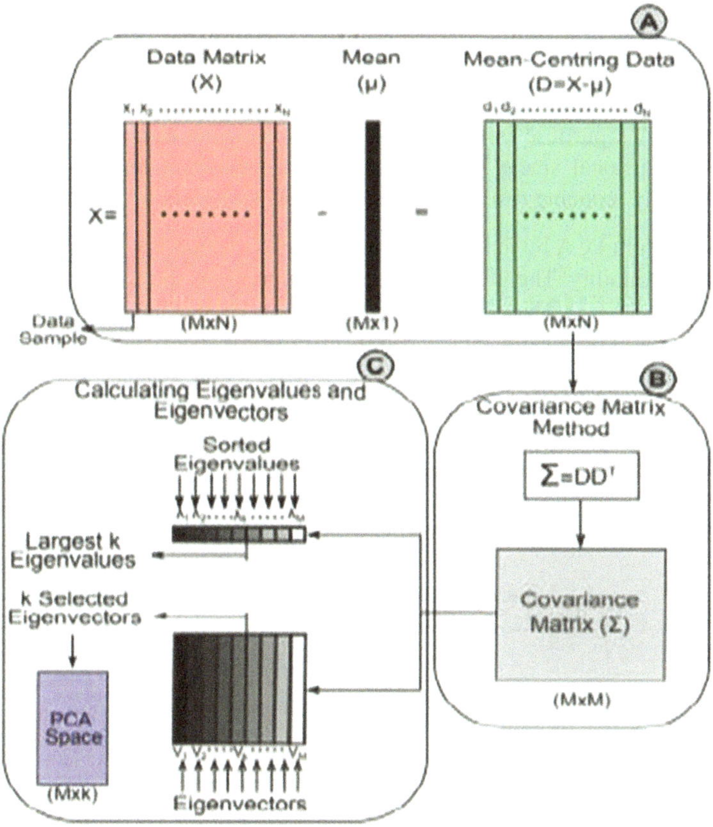

Fig. 2 Work flow for calculating the PCA based on covariance matrix [26]

The robustness of each component varies according to the amount of variance in its direction.

There are k principal components in the PCA dimension. The principal components are orthonormal and uncorrelated, and they represent the maximum variance direction.

The first principle component of the PCA space represents the direction of the data's biggest variance, the second principal component represents the second largest variance, and so on. As a result of its orthogonal principal component or axis rotation due to the rotation of the original axes, the PCA technique is considered an orthogonal transformation. The principal components can be calculated in two ways. The first process computes the covariance matrix, although the second uses the SVD method. Here we use covariance matrix to calculate the components. Figure 2 depicts how to calculate the PCA using covariance matrix.

Linear Discriminant Analysis

Linear discriminant analysis (LDA) is a linear regression model used to reduce the dimensionality of the given image space. LDA was introduced by Ronald A. Fisher in 1936 and is meant for two-dimensional and later for n-dimensional analysis. LDA's purpose is to project features in higher-dimensional space onto a lower-dimensional space in order to avoid the curse of dimensionality while simultaneously reducing resources and dimensional costs.

It differs from PCA by maximizing the inter-class variability and minimizing the intra-class variability. Therefore, LDA is used to measure the values within classes (s_w) and between (s_b). The advantage of maximizing the distance between the means of each class will project the data in lower-dimensional space, and it results in better classification accuracy. The working idea of LDA is that it projects the sample into a straight line with projection points as close as possible for inter-class and as widely away as possible for intra-class. When classifying the new sample, it should project onto the same line. The classification of this sample is determined based on the position of the projected point. The computations of scatter matrix within class (s_w) and scatter matrix between classes (s_b) are as follows:

$$s_w = \frac{1}{S}\sum_{c=1}^{S}\sum_{i=1}^{N_C}\left[\left(X_c^{(i)} - \overline{x}_c\right)\left(X_c^{(i)} - \overline{x}_c\right)^T\right] \tag{1}$$

$$s_b = \frac{1}{S}\sum_{c=1}^{S}\left(N_c\left(\overline{x}_c - \overline{x}\right)\left(\left(\overline{x}_c - \overline{x}\right)^T\right)\right) \tag{2}$$

s is variable which denotes the scatter matrix within in the class, while s_b denotes the scatter matrix between classes.

$X_c^{(i)}$, is finger knuckle image i of class c where c(c = 1, 2, 3, ... S), mean vector of class c is $\overline{x}_c$, overall mean vector is $\overline{x}$, and sample size of class c is represented by N_C.

The fisher criterion to measure the discriminatory power is calculated using Eq. (3):

$$J(w) = \frac{\left\| W^T S_b W \right\|}{\left\| W^T S_w W \right\|} \tag{3}$$

LDA will determine an orientation that allows the projected images to be separated well.

Given the following linear projection w,

$$y_i = w^T x_i \tag{4}$$

The y_i is a lower-dimensional feature vector for the image x_i.

3.2 Complex Feature Vector Based on PCA and LDA Parallel Fusion

The parallel fusion is combining the complex feature vector instead of super vector. Let us assume P and L are the feature vectors extracted from same finger knuckle sample F. In parallel combination the feature vector is represented as $F = P + iL$ (i is an imaginary unit). If the dimension of the P and L varies; then padding is done for lower-dimensional vector. For example, $P = (p_1, p_2, p_3)^T$ and $L = (l_1, l_2)^T$, padding is added for L and is represented as $L = (l_1, l_2, 0)^T$, and the parallel combination is denoted by $F = (p_1 + il_1, p_2 + il_{21}, p_{31} + i0)^T$.

3.3 Euclidean Distance Measure

Various distance functions have been used in the literature to measure the distance between points A and B in a feature space, with the Euclidean distance function being the most widely used. Let P and L be represented by feature vectors $P = (p_1, p_2, .., p_d)$ and $L = (l_1, l_2, .., l_d)$, where d is the feature space's dimensionality. The normalized Euclidean metric is commonly used to calculate the distance between P and L:

$$distance\,(P.\ L) = \sqrt{\sum_{i=1}^{d} \frac{(p_i - l_i)^2}{d}} \tag{5}$$

4 Results and Discussion

4.1 Data Description

PolyU Database

This section describes the PolyU Finger Knuckle Database, which is used for experimental analysis. Finger knuckle images were collected in a peg-free environment using a low-resolution camera from 165 people in the PolyU Finger Knuckle Print Database, with 125 males and 40 females. The images are captured in contactless mode with a resolution of 110 × 220 in BMP format. The images were collected in two different sessions. The average time between two finger knuckle capturing system sessions is thought to be 25 days. This database provides ROI images of captured finger knuckle prints by establishing a coordinate system for each FKP image [27]. The ROI image is extracted from the FKP image and contains the most significant feature that can be used for personal authentication. Some of the finger knuckle samples in database are shown in Fig. 3.

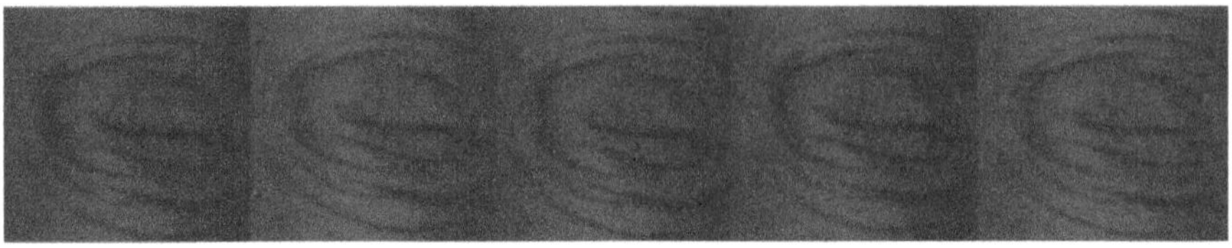

Fig. 3 PolyU FKP sample images

Fig. 4 IIT FKP sample images

IIT Delhi Database

The IIT Delhi Finger Knuckle Database, which is utilized for experimental analysis, is described in this section. The images were obtained in a contactless manner using a low-resolution digital camera. It contains 158 images of users' finger knuckles, ranging in age from 16 to 55 years. In total, 790 images of finger knuckles are available in the database [28]. Figure 4 shows some of the finger knuckle samples in the database.

5 Results

The primary goal of this experimental investigation is to determine the significance of single-finger dorsal surface patterns in the reliable biometric authentication process. Furthermore, this experimental study determines the efficiency of three proposed PCA_LDA CC techniques, namely, KNN, Euclidean, and neural network classifiers, which are implemented using two datasets, namely, (i) IIT Delhi dataset and (ii) PolyU dataset.

As a result, the performance evaluation for the three suggested PCA_LDA CC is carried out in two steps, namely, (i) Experiment 1 is carried out by comparing the PCA LDA with the PCA_LDA CC, i.e., serial fusing of vector with parallel fusion and (ii) Experiment 2 is carried out dependent on the number of classes.

5.1 Experiment 1

On two datasets, PolyU and IIT Delhi, the performance of the user authentication approach described in the previous sections was evaluated. The proposed system

Table 1 Analysis of FKP classification accuracy

Feature extraction	Accuracy (%)					
	IIT Delhi DB			PolyU DB		
	KNN	Euclidean	NN	KNN	Euclidean	NN
PCA_LDA	91.23	76.2	73.2	84.2	94.2	82.4
Proposed PCA_LDA CC	92.3	88.89	75	97.14	96.67	86.67

Table 2 Performance measures of PCA_LDA CC based on Euclidean classifier

No, of classes	Accuracy	
	IIT Delhi DB	PolyU DB
5	100	100
10	86.7	100
15	84.2	100
20	80.2	100
25	78.67	96
30	88.89	96.67

uses hybrid PCA_LDA complex conjugate features to recognize the FKP images. Most of the literature study shows that fusion takes place in serial. Here, the parallel fusion based on complex conjugate vector is proposed. These vectors are given to three classifiers, and results are shown in Table 1. The performance of the proposed feature extraction approaches is evaluated by three classifiers such as KNN, Euclidean, and neural network. The studies were carried out in order to identify the optimal performance of the finger knuckle print recognition system. The accuracy metric is used to calculate the FKP recognition performance.

5.2 Experiment 2

The other way of experiment is carried out to estimate the performance of the suggested PCA_LDA CC technique dependent on the number of classes (users).

Euclidean classifier is fed the proposed PCA_LDA CC (complex conjugate) features of the FKP image. The outcomes are depicted in Table 1. The accuracy of the PCA_LDA CC features' performance is evaluated using both IIT and PolyU FKP images. Each class has five FKP images, three for training and the remaining two for testing. Finally, the experimental outcomes demonstrate that the performance rates of 30 classes are 88.9% for IIT database and 96.67% for PolyU database, respectively.

KNN classifier is fed the proposed PCA_LDA CC (complex conjugate) features of the FKP. Table 2 displays the performance results. The accuracy of the PCA_LDA CC features' performance is evaluated using both IIT and PolyU FKP images. Finally, the experimental outcomes demonstrate that the performance rates of 30 classes are 92.3% for IIT database and 97.14% for PolyU database, respectively.

Table 3 Performance measures of PCA_LDA CC based on KNN classifier

No. of classes	Accuracy	
	IIT Delhi DB	PolyU DB
5	97	100
10	98	100
15	97.2	100
20	94.2	96
25	92.3	96.67
30	92.3	97.14

Table 4 Performance measures of PCA_LDA CC based on neural network classifier

No. of classes	Accuracy (%)							
	IIT Delhi DB Epochs				PolyU DB Epochs			
1	25	50	100	1	25	50	100	
5	86.6	63.3	86.6	86.6	93.3	93.3	93.3	100
10	71.2	70	70.1	81.1	83.3	100	90	93.3
15	75.6	62.33	66.7	64.44	86.7	85	80	95.6
20	58.6	68.3	61.67	68.36	86.6	83.3	85	78.3
25	68	62.74	61.33	60	80	84	81.3	87.6
30	74.29	68.10	65	78.1	86.7	86.7	81.1	87.6

Neural network classifier is fed the proposed PCA_LDA CC (complex conjugate) features of the FKP, and the result is shown in Table 3. The accuracy of the PCA_LDA CC features' performance is evaluated using both IIT and Poly U FKP images. The outcomes are studied in this section based on the number of epochs used, which can be 1, 25, 50, or 100. The feedforward network was utilized, with 60 hidden layers. Finally, the results validate that the performance rates of 30 classes for 100 epochs are 78.1% for IIT database and 87.6% for PolyU database, respectively.

5.3 Performance Comparison with Other Methods

Table 5 compares the performance of various other methods to the performance of the FKP proposed feature extraction techniques. The features of FKP are extracted using a hybrid PCA LDA with complex conjugate in the FKP proposed system. The proposed system's performance accuracy, which is based on three classifiers, is investigated. The PCA_LDA CC with KNN achieves a classification accuracy of 97.14%, which is higher than the 86.6% and 96.76% achieved by the neural network and Euclidean classifiers, respectively.

Table 5 Comparative investigation of proposed method with other approaches

References	Algorithm	Accuracy
Zhang et al. (2009	Competitive code [29]	76.26
Gao et al. (2014)	MoriCode [30]	74.79
Zhang et al. (2010)	ImCompCode [31]	84.05
El-Tarhouni et al. (2014)	Multi-scale LBP and PCA [32]	94.8
Zhang et al. (2010)	ImCompCode and Mag-Code [31]	87.43
Khellat et al. (2016)	FKP with SVM [33]	91.51
Zhai (2018)	Batch-normalized CNN [34]	90.05
Zhai (2018)	AlexNet [34]	85.6
Jaswal et al. (2017)	Deep matching of features [35]	99.10
Fei et al. (2019)	DDBFL [36]	92.21
Thapar et al. (2019)	FKIMNet [37]	94.02
Anbari and Fotouhi (2021)	Relaxed local ternary pattern (RLTP) [38]	96.43
Proposed method	PCA_LDA CC with NN	86.6
Proposed method	PCA_LDA CC with Euclidean	96.67
Proposed method	PCA_LDA CC with KNN	97.14

6 Conclusion

In this chapter finger knuckle recognition with parallel fusion of feature vector is proposed. The contribution of this work is the dimensionality of the FKP features which are reduced by extracting the features using the subspace algorithm such as PCA and LDA. Second is combining the PCA_LDA features based on parallel fusion. The features are extracted as complex conjugate feature vector, i.e., real and imaginary part of features, and this feature vector is combined based on parallel fusion. Third are the three classifiers such as Euclidean, KNN, and neural network to analyze the performance of the finger knuckle print recognition of the proposed feature extraction approach. The experimental results are conducted on two available FKP databases such IIT Delhi and PolyU. The results reveal that the suggested FKP recognition system out-performs other approaches in terms of accuracy. The knuckle print of a finger can be employed in real-time applications.

References

1. R. Clarke, Human identification in information systems: Management challenges and public policy issues. Inf. Technol. People **7**(4), 6–37 (1994)
2. S. Barra, M. De Marsico, M. Nappi, F. Narducci, D. Riccio, A hand-based biometric system in visible light for mobile environments. Inf. Sci. **479**, 472–485 (2019)
3. K. Usha, M. Ezhilarasan, Finger knuckle biometrics–a review. Comput. Electr. Eng. **45**, 249–259 (2015)
4. G. Jaswal, A. Kaul, R. Nath, Knuckle print biometrics and fusion schemes overview, challenges, and solutions. ACM Comput. Surv. **49**(2), 1–46 (2016)

5. K. Delac, M. Grgic, S. Grgic, Independent comparative study of PCA, ICA, and LDA on the FERET data set. Int. J. Imaging Syst. Technol. **15**(5), 252–260 (2005)
6. S.N. Borade, R.R. Deshmukh, S. Ramu, Face recognition using fusion of PCA and LDA: Borda count approach, in *24th Mediterranean Conference on Control and Automation (MED)*, (IEEE, 2016), pp. 1164–1167
7. G.L. Marcialis, F. Roli, Fusion of LDA and PCA for face verification, in *International Workshop on Biometric Authentication 2002*, (Springer, Berlin, Heidelberg, 2002), pp. 30–37
8. T. Peli, M. Young, R. Knox, K. K. Ellis, F. Bennett, Feature-level sensor fusion, in *Sensor Fusion: Architectures, Algorithms, and Applications III* 3719, (1999), pp. 332–339
9. C. Liu, H. Wechsler, A shape-and texture-based enhanced fisher classifier for face recognition. IEEE Trans. Image Process. **10**(4), 598–608 (2001)
10. D.L. Woodard, P.J. Flynn, Finger surface as a biometric identifier. Comput. Vis. Image Underst. **100**(3), 357–384 (2005)
11. A. Kumar, C. Ravikanth, Personal authentication using finger knuckle surface. IEEE Trans. Inf. Forensics Security. **4**(1), 98–110 (2009)
12. M.R. Swati, M. Ravishankar, Finger knuckle print recognition based on Gabor feature and KPCA+LDA, in *International Conference on Emerging Trends in Communication, Control, Signal Processing and Computing Applications (C2SPCA)*, (IEEE, 2013), pp. 1–5
13. S. Shariatmadar, K. Faez, A novel approach for finger-knuckle-print recognition based on Gabor feature fusion, in *4th International Congress on Image and Signal Processing 2011*, (2011), pp. 1480–1484
14. N.E. Chalabi, A. Attia, A. Bouziane, Multimodal finger dorsal knuckle major and minor print recognition system based on PCAnet deep learning. ICTACT J. Image Video Process. **10**(3), 2153–2158 (2020)
15. J. Kim, K. Oh, B.-S. Oh, Z. Lin, K.-A. Toh, A line feature extraction method for finger-knuckle-print verification. Cogn. Comput. **11**(1), 50–70 (2019)
16. M. Turk, A. Pentland, Eigenfaces for recognition. J. Cogn. Neurosci. **3**(1), 71–86 (1991)
17. P.N. Belhumeur, J.P. Hespanha, D.J. Kriegman, Eigenfaces vs. fisherfaces: Recognition using class specific linear projection. IEEE Trans. Pattern Anal. Mach. Intell. **19**(7), 711–720 (1997)
18. J. Li, S. Fong, R.K. Wong, V.W. Chu, Adaptive multi-objective swarm fusion for imbalanced data classification. Inf. Fusion **39**, 1–24 (2018)
19. Y. Chen, W. Xu, J. Zuo, K. Yang, The fire recognition algorithm using dynamic feature fusion and IV-SVM classifier. Clust. Comput. **22**(3), 7665–7675 (2019)
20. L.M. Abualigah, A.T. Khader, E.S. Hanandeh, A new feature selection method to improve the document clustering using particle swarm optimization algorithm. J. Comput. Sci. **25**, 456–466 (2018)
21. K. Zheng, X. Wang, Feature selection method with joint maximal information entropy between features and class. Pattern Recogn. **77**, 20–29 (2018)
22. H. Leng, X. Li, J. Zhu, H. Tang, Z. Zhang, N. Ghadimi, A new wind power prediction method based on ridgelet transforms, hybrid feature selection and closed-loop forecasting. Adv. Eng. Inform. **36**, 20–30 (2018)
23. J. Cai, J. Luo, S. Wang, S. Yang, Feature selection in machine learning: A new perspective. Neurocomputing **300**, 70–79 (2018)
24. J. Zhang, Y. Wu, W. Feng, J. Wang, Spatially attentive visual tracking using multi-model adaptive response fusion. IEEE Access **7**, 83873–83887 (2019)
25. S. Gu, R. Cheng, Y. Jin, Feature selection for high-dimensional classification using a competitive swarm optimizer. Soft. Comput. **22**(3), 811–822 (2018)
26. A. Tharwat, Principal component analysis-a tutorial. Int. J. Appl. Pattern Recognit **3**(3), 197–240 (2016)
27. http://www4.comp.polyu.edu.hk/csajaykr/fn1.htm
28. http://www4.comp.polyu.edu.hk/csajaykr/knuckle/iitdknuckle.htm
29. L. Zhang, L. Zhang, D. Zhang, Finger-knuckle-print: A new biometric identifier, in *Proceedings of the 16th IEEE International Conference on Image Processing (ICIP)*, (2009), pp. 1981–1984

30. G. Gao, J. Yang, J. Qian, L. Zhang, Integration of multiple orientation and texture information for finger-knuckle-print verification. Neurocomputing **135**, 180–191 (2014)
31. L. Zhang, L. Zhang, D. Zhang, H. Zhu, Online finger-knuckle print verification for personal authentication. Pattern Recogn. **43**(7), 2560–2571 (2010)
32. W. El-Tarhouni, M.K. Shaikh, L. Boubchir, A. Bouridane, Multi-scale shift local binary pattern based-descriptor for finger-knuckle-print recognition, in *2014 26th International Conference on Microelectronics (ICM)*, (IEEE, 2014), pp. 184–187
33. S. Khellat-Kihel, R. Abrishambaf, J.L. Monteiro, M. Benyettou, Multimodal fusion of the finger vein, fingerprint and the finger-knuckle-print using kernel fisher analysis. Appl. Soft Comput. **42**, 439–447 (2016)
34. Y. Zhai, H. Cao, L. Cao, H. Ma, J. Gan, J. Zeng, et al., A novel finger-knuckle-print recognition based on batch-normalized CNN, in *Chinese Conference on Biometric Recognition*, (Springer, Cham, 2018), pp. 11–21
35. G. Jaswal, A. Nigam, R. Nath, Finger knuckle image based personal authentication using deep matching, in *IEEE International Conference on Identity, Security and Behavior Analysis (ISBA)*, (IEEE, 2017, February), pp. 1–8
36. L. Fei, B. Zhang, S. Teng, A. Zeng, C. Tian, W. Zhang, Learning discriminative Finger-Knuckle-print descriptor. IEEE Int. Conf. Acoust, Speech Signal Process. (ICASSP), pp. 2137–2141 (2019)
37. D. Thapar, G. Jaswal, A. Nigam, FKIMNet a finger dorsal image matching network comparing component (major, minor and nail) matching with holistic (Finger Dorsal) matching. Int. Joint Conf. Neural Netw. 1–8 (2019)
38. M. Anbari, A.M. Fotouhi, Finger knuckle print recognition for personal authentication based on relaxed local ternary pattern in an effective learning framework. Mach. Vis. Appl. **32**, 55 (2021)

A Systematic Literature Review on Data Freshness for Reinforcing Mutual Authentication in Wireless Body Area Networks

S. Aarthi, K. Geetha, N. Sasikaladevi, and C. Mala

Abstract With the recent improvements in wireless communication technologies, the Wireless Body Area Network initiates various applications to fulfill the people's requirements in healthcare. WBAN is a reliable ad hoc communication that deals with the physiological information of patients by providing better real-time health monitoring services. WBAN deals with the sensitive physiological information collected by the biomedical sensors distributed over the human body, and these observations are exchanged to the healthcare provider through the Internet without affecting the user's comforts. Data freshness ensures the freshness of cryptographic parameters passed during the mutual authentication process of WBAN. Achieving data freshness includes various challenges in WBAN due to its resource constraints and limited applicability. In this work, a comprehensive study of the WBAN design challenges with communication architecture is presented, with a prime focus on the data freshness to enhance the security properties accompanied by suggested research directions, and enhancements are exhibited.

Keywords Wireless Body Area Network (WBAN) · Data freshness · Resource constraints · Mutual authentication · Real-time health monitoring services

1 Introduction

As reported by the International Diabetes Federation in 2020, it has been confirmed that more than 463 million people are affected by diabetes worldwide, among which 77 million people belong to India; this count is estimated to increase by 2045, and

S. Aarthi (✉) · K. Geetha · N. Sasikaladevi
CSED, SoC, SASTRA Deemed University, Thanjavur, Tamil Nadu, India
e-mail: aarthiresearch047@gmail.com; geethavalavan@gmail.com; sasikalade@gmail.com

C. Mala
Department of Computer Science and Engineering, National Institute of Technology, Tiruchirappalli, Tamil Nadu, India
e-mail: mala@nitt.edu

A. Haldorai et al. (eds.), *4th EAI International Conference on Big Data Innovation for Sustainable Cognitive Computing*, EAI/Springer Innovations in Communication and Computing, https://doi.org/10.1007/978-3-031-07654-1_14

there will become 134 million confirmed diabetic cases to be reported in India alone. Additionally, based on the statistics estimated by the World Health Organization in 2019, 17.9 million people lose their lives due to cardiovascular diseases each year. Out of these deaths, one third of the people are under the age of 60 years. As a result of considering these analytical reports, the total expenses covered for the conventional healthcare monitoring systems are expected to increase exponentially, and the people require the real-time healthcare assessment with enhanced usability. These limitations in the health systems stimulate the dominance of Wireless Body Area Network (WBAN) to improvise the healthcare monitoring procedures without affecting the quality of the patient's lives. WBAN is required to monitor the health conditions of the patient by capturing the vital signs such as heart rate, respiratory level, blood pressure, and ECG and body temperature of the patient as a real-time observation or data.

In real time, the medical professional accesses the real-time observation of the patient through the hub node or local server. Data access over the insecure channel forces the security analyst to develop the better authentication protocol. Otherwise it will lead to various threats and vulnerabilities that results in misleading intervention of the medical professional [1]. Data confidentiality and privacy are the prime requisites for data transmission in WBAN because it handles sensitive information concerning the health conditions of the patients. To ensure authenticity in resource-constrained environment, many security analysts designed various authentication protocols to fulfill the security goals [2], [3], [4], [5], [6], and [7]. Any compromise would not only affect the patient's privacy, but it may also endanger the lives of the patient. Data freshness is defined as the security property to ensure that the cryptographic parameters received during the mutual authentication process [8] are not replayed; only the current and updated data is received at the receiver's end. This systematic study focuses on the novel diversions of data freshness mechanisms in smart healthcare applications. The main objective is to concentrate on the implementation of data freshness mechanisms in various authentication and key exchange protocols in WBAN. Towards exploring the results, we will identify, plan, and conduct the review on existing works and discuss the results to formulate the new research directions.

1.1 Motivation and Contribution

According to a data breach investigation report by Verizon, the data breaches in the health sector increased by 58% in 2021 when compared with the last year. For instance, the false respiration rate sent to the medical analyst may lead to improper consequences in critical situations due to the misleading treatments of the medical professional. Similarly, in the case of military applications, a data breach may endanger the lives of the battalion in the war field. In WBAN, numerous security vulnerabilities and threats are associated with it. So, data freshness is the most prominent security requirements to ensure the freshness of the cryptographic parameters passed during the mutual authentication protocol. This SLR aims to provide a brief summary about various data freshness mechanism used in Wireless Body Area Network. The cryptographic parameters like identity of the node, timestamps, hash functions, hashed parameters, and random numbers must be fresh to establish a unique session key. If this property fails, an adversary can compromise the entire network. In WBAN, data freshness is mainly utilized to assure the replay protection for the data passed over an insecure public channel. In this review, we concentrate on 1977 journals of WBAN security from various research platforms. We conduct a review on 36 chosen research articles refined from the primary and the secondary search process related to data freshness and authentication mechanisms in WBAN. In this SLR, we analyze WBAN communication architecture and data freshness mechanisms related to the mutual authentication process.

1.2 Paper Organization

The rest of the paper is described as follows. Section 2 demonstrates the WBAN communication architecture and session key establishment in IEEE 802.15.6 to improve the readability of the paper. Section 3 suggests the research methodologies including the selection results, inclusion and exclusion principles, and quality assessment for this systematic review process. Section 4 describes the results of the framed question as the research goals and relates how the data freshness plays a vital role in mutual authentication process of WBAN. Section 5 concludes the various approaches of data freshness with suggested future directions.

2 Preliminaries

2.1 WBAN Communication Architecture

For understanding the security issues in Wireless Body Area Networks, it might be significant to investigate the structure of the WBAN communication. The classification of tiers depends on whether the communication takes place between the internal entities of WBAN or outside the WBAN. In this section, multiple communication tiers present in WBAN are discussed. Figure 1 illustrates the nodes disseminated by the specific distance in the network based on its applicability. The network topology of the WBAN is categorized depending on the location and characteristics of the nodes. They are *Simple single hop network topology* [9] in which the sensor nodes and hub nodes are in the direct communication range, and no need to use the relay nodes. The latter is a *centralized two-hop star topology* in which the sensor and hub node is not in the direct communication range. For improving network efficiency, relay nodes are used to provide a two-hop communication between the sensor and

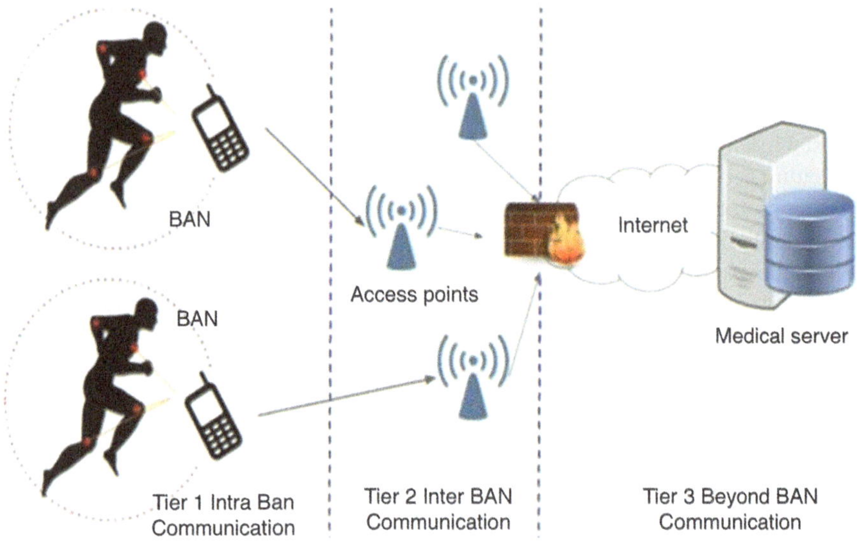

Fig. 1 Architecture of a WBAN system

hub node. Before discussing the communication architecture, we need to impart knowledge over the nodes and their usability in WBAN.

Sensor Nodes (End Nodes) Sensor nodes are the hardware peripheral units that are used to collect and exchange physiological information to the local server. Each sensor has the ability and properties to observe the specific biological readings of the patient. In the WBAN architecture, they are interpreted as end nodes.

Relay Node (Intermediate Nodes) If the sensors and hub nodes are not in the proper network communication range, these are the intermediate nodes to provide enhanced coverage in WBAN. These nodes are responsible for the process of relaying in the two-hop star topology of WBAN.

Hub Node (Local Server) These nodes exchange information between the sensor and the medical advisor. This hub node links with all the other nodes present inside and outside the Wireless Body Area Network.

The above insights provide the needed information about the nodes and their capabilities. Next, we need to shift the attention over the communication tiers of WBAN. The various tiers of communication are encompassed as follows:

Tier I Intra-WBAN Communication In this tier, the interaction among the sensor nodes is restricted around the body. The communication among the end nodes and the interaction between the end nodes and relay nodes constitute the tier I communication that helps to transfer the observations to the next tier of communication. Tier I communication is handled by low-energy protocols including the LWPAN, Bluetooth, Wi-Fi, and cellular to avoid resource constraints.

Tier II Inter-WBAN Communication The communication between the relay nodes and hub nodes comprises the tier II communication in WBAN. This type of communication is allowed between the coexisting WBAN inside the permissible network. So, the information is retrieved and accessed through various mediums with the help of the Internet.

Tier III Beyond WBAN Communication In this tier, the hub node transmits the collected observations about the patient to the medical advisor. Finally, the medical advisor analyzes eHealth reports of the patient and alerts the patient in case of emergencies. This communication takes place between the hub node and the medical server. Although the WBAN is equipped with simplicity and easement, it suffers a lot due to the low security and resource constraints. In terms of security, the prime objective of WBAN is to establish the above three-tier communication with the following security services. They are confidentiality, authenticity, and integrity.

2.2 Session Key Establishment

In IEEE 802.15.6 standard, the *security association* is used to uniquely identify the sensor node and local server together. The security association is generated at the network layer by validating the authentication credentials of the user. The system administrator verifies the identity of each node before providing the master key for key establishment [6]. System admin at the medical server-side shares the master key or pre-shared key among the sensor node and local server to establish a new session key for further communication. There are different methodologies used to generate the session key. They are key transport protocol, key agreement protocol, and the hybrid protocol for authenticated key exchange. If any of the protocol parties create the session key and it is distributed to all the other parties present in the communication, it is described as the key transport protocol. If all the protocol parties provide some valuable input as a parameter to generate the session key, it is called key agreement protocol. Each has its beneficiaries and disadvantages. In practice, by verifying the identity of the sensor and hub node, the system admin provides a master key for the security association that can be used once between the two parties [10]. After performing some cryptographic methods, a pairwise temporal key is generated from the master key once per session.

Figure 2 explains the session key establishment as a result of the mutual authentication in Wireless Body Area Network. Authentication credentials may have various static and dynamic attributes to generate the master key from the system administrator employed in medical server [11]. This master key provides the pairwise temporal key between the protocol entities, and this pairwise temporal key is used for providing security properties in WBAN using the encryption,

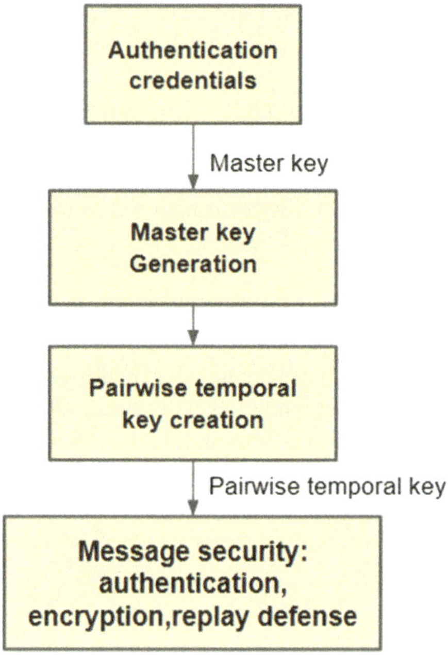

Fig. 2 Session key establishment in IEEE 802.15.6

authentication, and replay protection mechanisms. Among these processes, replay protection is established using the data freshness [12]. In the field of IoT, researchers proposed various authentication schemes like symmetric-based schemes, biometric-based schemes, public key signature schemes, and block chain-based schemes discussed in Sect. 4.

3 Research Methodology

This section provides the preliminary steps followed in primary and secondary search phases to refine the research articles in the field of mutual authentication in WBAN. To focus on the research questions, the SLR conducts a review of refined existing research studies on data freshness mechanisms and security in WBAN.

3.1 Research Goals

The main benefit of this research is to investigate the risk factors in existing research studies and for summarizing the efforts and advancements of research in data freshness schemes utilized in resource-constrained environments like WBAN. For conducting the SLR in a planned way, we framed four research questions tabulated in Table 1.

Table 1 Research questions

S. no	Research questions	Significance
RQ1	What are the challenges in exhibiting the data freshness mechanism?	It concerns the processing capability of biomedical sensor nodes, the requirement of energy, security, and privacy. It will help to improve the knowledge over the general challenges and converge into the challenges for developing the data freshness mechanisms in WBAN
RQ2	What are the security vulnerabilities in WBAN?	Due to the resource constraints, WBAN is subjected to various risks and parametric limitations. This makes us to understand the threats and vulnerabilities in WBAN
RQ3	What kinds of mechanisms are defined for data freshness in WBAN?	This research question helps to understand the various mechanisms used for obtaining the data freshness
RQ4	What kind of methods is available to insist on data freshness to obtain authenticity through the lightweight cryptographic protocols?	This question will help for scrutinizing the effective methods for reinforcing the mutual authentication scheme with the data freshness property in WBAN

3.2 Selection of Research Articles

This section focuses on the selection of the research articles based on the keywords in primary search process. This primary search process is focused on looking for the keywords from a particular research platform that motivates this SLR. The keywords were chosen to provide the objective of answering the research questions in Table 1. The search strings which assist in this SLR were:

(Data freshness maintenance in WBAN OR freshness of data in WBAN OR security requirement analysis for WBAN OR WBAN security) AND Authentication in WBAN

These keywords were searched on research platforms like ACM Digital Library, IEEE Xplore Digital Library, SpringerLink, Google Scholar, and ScienceDirect. The primary search process was organized on 4 October 2021. Then these results were passed to the secondary search process which filtered the research articles based on the inclusion and exclusion criteria mentioned in Table 2.

Table 2 Inclusion and exclusion principles for the primary search process

S.no	Inclusion principles	Exclusion principles
1	The research article must illustrate the importance of data freshness in WBAN	General studies of WBAN and nontechnical papers and blog posts
2	The research article must possess the experimental discussions with formal and informal security analysis of the protocol	The research articles without formal verification of the protocols and do not discuss the importance of data freshness in a resource-constrained environment
3	The research article must be peer-reviewed and published in technical journals or conference proceedings	Non-English papers

3.3 Inclusion and Exclusion Criteria

The research studies to be selected must include the experimental discussions and be lightweight to suit the resource-constrained environment. The research articles must summarize the formal and informal security analysis of the protocol with technical advancements in the research directions. The research studies with these criteria were included for this review process. And general studies, nontechnical papers, and blogs about WBAN were strictly factored out in this review process. The inclusion and exclusion criteria were tabulated in Table 2.

3.4 Quality Assessment

This process allows evaluating how the selected research article will answer the formulated research question in Table 1. The assessment must determine the effectiveness and experimental findings of the paper in the field of data freshness and its application towards lightweight cryptography in WBAN. After assessing the refined 57 research articles, only 36 research articles are passed in the quality assessment process. Now the review process focused on 36 research articles concerned with the session key establishment protocol with mathematical proof and algorithmic construction. Focusing on these research articles, a particular important challenge is the assurance of replay-resistant mutual authentication protocols in WBAN. Thus this SLR achieves the comprehensive study of various data freshness approaches in key management phase in this resource-constrained environments.

4 Results of the Systematic Literature Review

This section deals with the empirical results and findings of the conducted systematic literature review on WBAN. The results were discussed to answer the research questions of security and privacy in resource-constrained environments. This section can be subdivided into four subsections. Each subsection deals with the individual research questions and formulation of the objectives to answer the research question with tabulation and charts over the existing methodologies.

4.1 RQ1: What Are the Challenges in Exhibiting the Data Freshness Used to Ensure Authenticity in WBAN?

General Challenges of WBAN WBAN comprises various nodes connected through some wireless communication technologies. Based on the locality and performance of the node, the communication capability of the node differs. Due to the strictly resource-constrained devices, WBAN must provide reliable communication with latency and jitter in an acceptable range. For instance, latency should be less than 125 ms for medical applications and 250 ms for nonmedical applications. Considering jitter should be less than 50 ms [13], it must also design an energy-efficient mechanism for reducing the power consumption of biomedical sensor nodes. WBAN allows authentication, confidentiality, and integrity-related services which utilize lesser resources. WBAN must support recoverable methodologies in case of any node failures. Finally, it must extend to the heterogeneous wireless environment, and the data rate should be from 10Kbps to 10Mbps [14].

Security Challenges of WBAN WBAN deals with sensitive information. Any loss of data will not only affect the patient's privacy. It leads to the improper intervention of the medical professional and endangers the lives of the patient. To avoid this, WBAN must support the authentication and encryption mechanisms [15]. But substantiating the conventional cryptographic algorithm leads to higher memory and energy requirement, key size, and processing capability. These cryptographic primitives will not be applicable for resource-constrained WBAN. So, security challenges involve the storage requirements, computational overheads, and communication costs [16] of the security primitives.

4.2 RQ2: What Are the Security Vulnerabilities in WBAN?

Due to the resource constraints in devices like sensors and relay nodes in WBAN deployment, various attacks are possible in healthcare applications.

Eavesdropping Attack The passive adversary eavesdrops on the conversation between the medical professional and the patients using the insecure wireless channels. It will seriously impact the interception of sensitive information like ECG, pulse rate, and glucose level. Even though the information is well encrypted and authenticated, there will be a possibility of monitoring the conversation by the passive adversary. He may also try to extract the cryptographic parameters passed in mutual authentication protocol by snooping into the medium.

Replay Attack Adversary replays the past messages to gain control over the WBAN communication. The adversary sends the replayed messages based on the eavesdropped conversation between the patient and medical professional. It will cause serious confusion while treating the patients for chronic diseases.

Impersonation Attack Here the adversary impersonates a legitimate user for gaining more private sensitive information in the WBAN network. All the authorized entities may recognize the adversary as an authorized one who has past session keys and messages.

Man in the Middle Attack The adversary may intercept the conversation between the patient and medical professional. In this relay attack, the attacker gains information from both the client and medical professional, and he may also modify them.

Key Compromisation Attack Adversary acts as the various authorized entities in the WBAN network to produce a valid session key. This attack may breach the authentication, confidentiality, integrity, and data freshness properties. The adversary gains complete knowledge over the mutual authentication protocol that is utilized in the WBAN conversation.

Gateway Bypassing Attack In this attack, the adversary captures the relay node and feeds the past messages to bypass the authentication process.

Node Capture Attack The adversary captures the legitimate client of the WBAN and makes some conversation to compromise the entire network. This attack may lead to various malicious activities in the network without knowing the identity of the perpetrator.

These are the attacks that will induce adverse effects on the authentication process. While transmitting the cryptographic parameters for session key establishment, we must resist the above attacks by introducing the data freshness mechanisms. This data freshness property will enhance the integrity of the crypt parameters transmitted over the network.

4.3 RQ3: What Kinds of Mechanisms Are Used for Data Freshness in WBAN?

Data freshness is one of the security services which provide integrity and data protection for the cryptographic parameters passed during the mutual authentication protocols in WBAN. Furthermore, data freshness introduces the non-malleable property for any key exchange protocol used in IEEE 802.15.6. Data freshness is defined to verify whether the cryptographic parameters are current and not replayed from any other previous sessions. The workflow of the WBAN is illustrated in the Fig. 3, in which the authentication request and response must be protected with integrity and real-time data protection. Key agreement protocols like Diffie-Hellman share their valuable input values to produce the session key. For instance, the random number chosen by the sensor node and hub node can uniquely derive the session key with some cryptographic functions.

One noteworthy feature about this methodology is that the session key is derived only if all the protocol entities share their specific value. If not, the session key is not established and the communication will be interrupted. But the problem is we need to wait till all the entities share their specific input value and agree upon a key. It is more time-consuming. In the WBAN environment, we require high throughput and low latency because it deals with sensitive information and needs assessment within a specific amount of time. Otherwise the diagnosis of the patient will be delayed in emergency situations and even harm their lives.

Mechanisms to Guarantee the Data Freshness For providing high throughput and low latency in lightweight key agreement protocols, the sender has to send a message with some additional information to ensure the freshness of that message. In this case, one entity which is a system administrator needs to provide a key to all the other nodes such as sensor node, relay node, or hub node. Here nodes other than the system administrator do not have any part in session key generation. So,

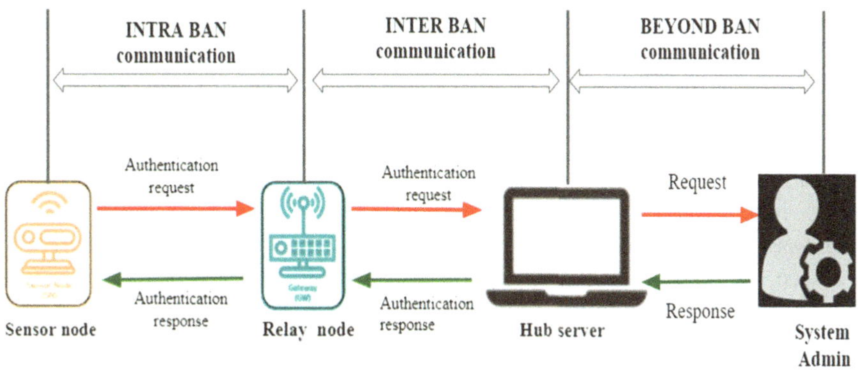

Fig. 3 The workflow of WBAN

they investigate whether the session key distributed from the system administrator is a current or replayed one. For ensuring the freshness of the data, the protocol entities need to check some parametric values associated with the messages. In this method, time-consuming is less, but only partial integrity is established. A typical example behind this method is data protection in Temporal Key Integrity Protocol (TKIP), in which the sequence counter is placed to denote the sequence number of each packet sent between the sender and receiver and the corresponding sequence number table maintained at both ends. If the sequence number is greater than the previous sequence number, then the packet will accept. These are the two different ways to guarantee the freshness of the communication.

Parametric Values for Data Freshness In practice, the parameters used for ensuring the freshness values are timestamps, random challenges, and sequence counters.

Timestamps When using this kind of approach, the sender has to append the current timestamps with the message and send them to the receiver. In the WBAN authentication phase [17], [18], [13] and [14], the sensor node appends the current timestamp with the identity information to create a temporary identity, and this timestamp is transmitted along with the information of the sensor node to exchange the valuable cryptographic parameters with the local server. Now the local server verifies the timestamp of the message received and checks its validity by calculating the difference between the current time and the timestamp provided by the sensor node. If the difference is negligible, the hub node accepts the information and establishes the connection with the sensor node. These procedures are used in WBAN to create the temporary identification information of the node to implement the anonymity property explained in Fig. 4. Among all the other methods, this parametric value provides an easier implementation. If the clock of the recipient is advanced than the current time due to some technical errors, then the recipient does not accept the messages sent by the sender. And another limitation is legitimate users sent the messages properly, but due to some communication delays, the data packets were received late. Here the legitimate messages from the authorized sender are not accepted by the recipient because of the invalid timestamp. The legitimate messages sent by the sensor node are captured by an adversary and delayed it on purpose. The ultimate goal of the adversary is to forge the legitimate messages as an invalid messages at the receiver's end. This is termed as suppress replay attack [19], [20], and [21].

Random Challenges This mechanism enforces the challenge-response model. For instance, the sensor node is connected with the mobile for three-factor authentication through Bluetooth protocol. In this phenomenon, the sensor node and mobile are already paired up with each other using authentication credentials. Now the mobile node tries to authenticate the sensor node again to check whether the link key is fresh or not. In this stage, a random number (nonce) [22] and [12] is chosen by the mobile node as an authentication request to the sensor node. Upon receiving

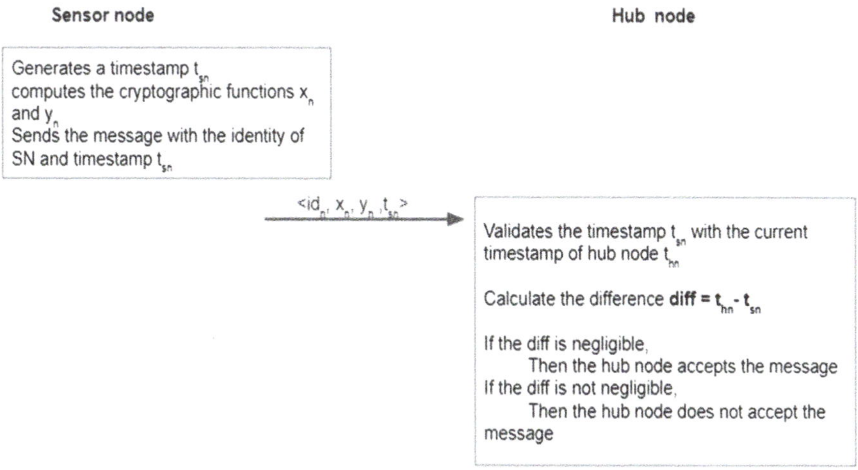

Fig. 4 Verifying data freshness using timestamps in key establishment

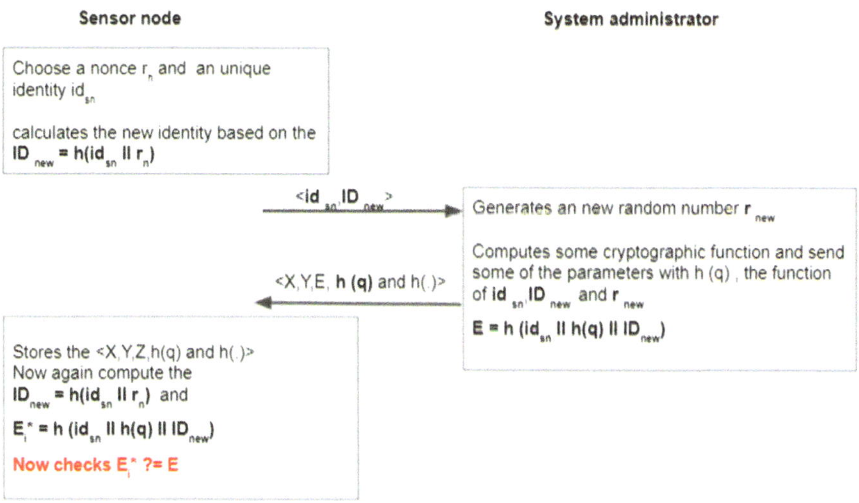

Fig. 5 Verifying data freshness of parameters using random challenges

the request, the sensor node generates the response as a function of the nonce, link key, and Bluetooth address of the sensor node. By comparing the authentication response, the mobile node verifies whether the key established between the sensor and mobile node is fresh or not (Fig. 5). The main disadvantage is the *pre-play attack* in which the adversary guesses the random number before the completion of the mutual authentication protocol. If the range of the random number is less, then it is easily predictable by an adversary.

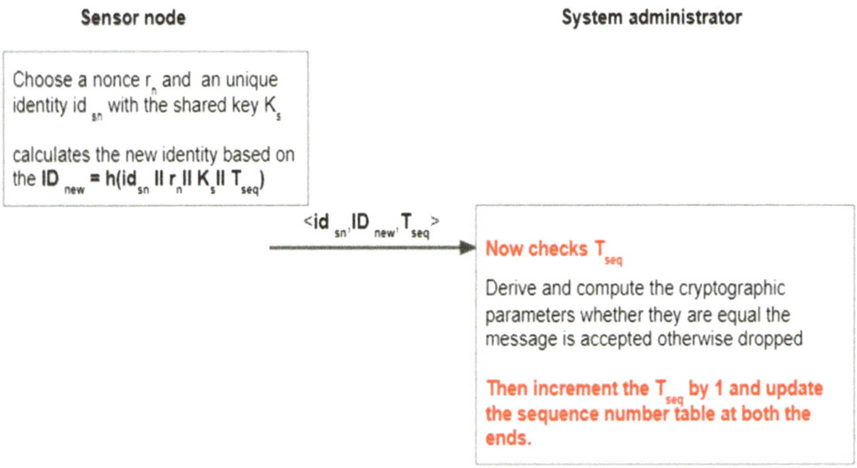

Fig. 6 Verifying data freshness of parameters using counters

Counters Whenever the message is sent by the sender, the sequence counter value must be appended with the message, and then the counter value is incremented by 1. For instance, in the Counter CBC MAC protocol (CCMP), the sequence number is replaced by the packet number as a count value. The current packet received must have a greater packet number than the already received packets. But TKIP and CCMP follow the same procedure to check whether the data is fresh or not. If the clock is not synchronized or contains some synchronization errors, it will lead to adverse effects and vulnerabilities in the replay protection mechanism (Fig. 6).

Hybrid Methods By using the above three parametric values, some protocols of AKE use a hybrid mechanism [23], [15], and [24] for the data freshness. The sensor node needs to pick a random number (nonce), and the current timestamp concatenates it with the message to be transmitted. Upon validating the timestamp and the random number, the hub node accepts the message. In the application of electronic payment system, the smart card user sends the message to the server with the current timestamp and nonces.

If the attacker captures the messages m1, m2, and m3 and sends them with the same timestamp as it is in the messages, then the server or user can easily identify the replayed messages by validating the timestamps. But if the adversary modifies the previous timestamps into the current one, then the adversary will act as a legitimate user. In this situation, timestamp verification is not only enough. So to avoid this, the author presented the method to verify the timestamp with a random number. Even if the adversary modifies the timestamp, he cannot identify the nonce generated by the authorized user. By providing this kind of method, replay protection is preserved in WBAN. The retransmission of the intercepted messages is resisted through the timestamp and random numbers in [22] and [3].

	Algorithm 1: Verifying data freshness of cryptographic parameters
	At sensor
Step 1	Generate an identity of the node $\mathbf{id_{sn}}$, a random number $\mathbf{r_n}$, and timestamp $\mathbf{t_{sn}}$ by the sensor node
Step 2	Computes the cryptographic parameters and temporary id using $\mathbf{tID = h(id_{sn} \oplus t_{sn} \| r_n)}$
Step 3	Send the cryptographic parameters $\mathbf{<tID, y, A, B, t_{sn}>}$ to the hub node or local server
	At hub node
Step 4	Checks the validity of the timestamp $\mathbf{t_{sn}}$ with $\mathbf{t_{hn}}$
Step 5	Derive the random number $\mathbf{r_{new}}$ based on the cryptographic parameters received
Step 6	Calculate the temporary id $\mathbf{tID_{new} = h(tID \oplus t_{sn} II\ r_{new})}$
Step 7	Checks the $\mathbf{tID_{new}?=tID}$ to ensure the cryptographic parameters are current one and not replayed from the previous sessions

4.4 RQ4: What Kinds of Methods Are Available to Insist on Data Freshness to Obtain Authenticity Through the Lightweight Cryptographic Protocols?

For securing the data communication, cryptographic mechanisms must indulge the end-to-end security among the protocol entities by excluding the interception of messages by unauthorized third parties. The communication cost and computational capability will be high to implement a better cryptographic mechanism for data transmission. But it is not possible in WBAN due to its resource constraints. So we need some other options to establish better cryptographic primitives and key exchange protocols which utilize lesser resources.

Biometric-Based Authentication Schemes Some of the existing biometric authentication and key agreement protocols, synchronous clocks are considered as the best solution [25], [26], and [27], because the dynamic biometric attribute is required for node authentication and transmission of data. If we use the static biometric attribute [19] like the face, iris, and fingerprint, then it will not provide resistance over the stolen key problem, privileged insider attack, and snooping attack. Xue et al. introduced a mutual authentication scheme through the smart cards, but it violates security goals and does not have resistant to stolen smart card attack and offline password guessing attack. But in dynamic biometrics, inter-pulse interval is taken as a most promising attribute. This IPI data is calculated from the two consecutive biological signals from ECG. Data freshness is ensured by calculating the hamming distance of the generated bio keys in IPI data. If the IPI data measured at the sender and the receiver is at different rate and time, then we need more computational analysis for that data. For this the data acquisition cost in both the sensor node and hub node will be high [25]. So we prefer the IPI data in real-time signals which need a synchronous clock for verifying the freshness of the dynamic biometric attribute. If any error is present in the received parameters, error-correcting code and morphing functions will be used to reconstruct the parameters. But the error must be in the acceptable range. Here, the problem is the usage of the error-correcting codes increases the computational complexity and communication overhead with storage requirements. Turkanoic et al. insisted the two-factor authentication protocol in WSN, but it surpasses the offline identity guessing attack, offline password guessing attack, and impersonation attack. To enhance this authentication, three-factor lightweight authentication protocol was suggested in [28] by Jiang et al. [29] using the bio-hashing methods. Das et al. proposed a biometric authentication scheme in WBAN, but the cost of computation is high.

Block Chain-Based Authentication Schemes These schemes [17], [30], and [30] suggested current timestamps need to be appended with the identity of the node to ensure the freshness of the data. Although it verifies the freshness of cryptographic parameters, these schemes do not provide much scalability in the dynamic node authentication phase. These schemes must be improved in the areas of hyper ledger fabric-based authentication schemes and distributed ledger security. Preferrably, timestamps are used for each and every authentication request among the protocol entities. But these schemes need high throughput and with negligible delay for each transaction carried in mutual authentication phase.

Symmetric-Based Schemes These schemes, [31] and [32], consider the pre-master key and the unique identifier of the nodes participating in the WBAN communication. For instance, the sensor node combines its unique identity with the current timestamp or random number to mask its identity using a temporary identifier for a particular session to ensure the properties like unlinkability and anonymity. Wong et al. proposed a novel lightweight PAKE (password-based authentication and key exchange) protocol for accessing the WBAN in which the random number is used to generate the data freshness for session key establishment phase [10]. In 2013, He et al. introduced the lightweight authentication using cryptographic and component-wise addition modulo 2 operations to resolve the issues in the resource-constrained devices [33] with timestamps for data freshness. This approach is a threat towards the forgery attack and man in the middle attack. Das et al. suggested the two-factor authentication scheme [34] with timestamps to ensure the freshness of cryptographic parameter results in various attacks like user impersonation attack, gateway node bypassing attack, and privileged insider attack and does not support mutual authentication process discussed in [35]. In 2015, Gope et al. designed the symmetric authentication scheme for resource-constrained devices with sequence counters as the solution for assuring the freshness of the data. But it is susceptible to the session key disclosure. Li et al. [23] proposed a novel authentication scheme for WBAN with random numbers and timestamps to ensure the freshness of the cryptographic parameters. But it does not become resilient over sensor node impersonation attack and loss in forward secrecy. Kompara et al. eliminated the drawbacks in Li et al. authentication scheme by keeping the synchronization in all the protocol entities participating in WBAN communication. But this lacks the forward secrecy. In 2020, Gupta et al. introduced provably anonymous lightweight authentication protocol by providing security over the gateway node authentication request and responses [36]. Mehmood et al. [9] considered the rekeying method in symmetric-based encryption schemes to increase randomness of the parameters in the authentication request and responses. Three-factor mutual authentication scheme is introduced in [37] and also ensures the freshness through the random number and timestamps.

Public Key Signature Schemes In these schemes, the public key signature is formalized by the two nonidentical keys. But the conventional public key cryptography is not effective in resource-constrained environment. The main limitation of PKC is the higher communication and computation cost. For instance, RSA signature is based on the congruent theory which implies the data freshness indirectly in the authentication protocol. If both the protocol entities share their value, the session key is generated. But the key length is 12 times higher than the symmetric crypto primitives. So to avoid the larger key length, elliptic curve cryptography primitive is an emerging field with higher efficiency than any other asymmetric crypto primitives. But the cost of operation is higher when comparing with the symmetric cryptosystems or the cryptosystems with pre-deployed keys. And it is compromised using the key compromise impersonation (KCI) attack and lacks the forward secrecy policy. In public key cryptographic schemes, the signer requires the digital certificate from the certificate authority to bind with the message to be verified by the verifier to ensure authenticity. This certificate itself acts as a parameter of data freshness. But managing the certificates and processing the certificates through the certificate authority increase the complexity over communication and storage requirements. Finally, these attacks are possible only if the public key signature schemes are not properly validated.

ID-Based Signature Schemes For avoiding the limitations of PKI, identity of the user needs to be bound with the signer's public key to generate the session key. Identity-based signature schemes [38], [20] do not require any digital certificate from the hub node and system administrator to authenticate each other. In authentication, trusted third party must generate the master time key and store it for validating the timestamps of the messages of the signer and verifier. Then TTP transmits the unique identifier for the hub node (signer) to establish the session by sending identity of the signer and timestamp and signing key to the system administrator (verifier). Here, the verifier verifies the freshness of the message and signature by validating the timestamp sent by the signer and the current timestamp at the side of the system administrator. But it needs the trusted third party (TTP) to authenticate them. If the trusted third party is compromised by an adversary, then the adversary can even able to decrypt the past communication and invade the structure of the key establishment protocol . This scheme does not become resilient over key escrow problem.

Certificateless Signature Schemes These schemes [39] eliminate the key escrow problem by enriching the security goals. Liu et al. suggested a certificateless efficient authentication scheme in WBAN based on the short certificateless digital signature method. In this scheme timestamp is used to provide anonymity and unlinkability for the protocol entities. But for node authentication, timestamp value and identity of the node is easily traced out by an adversary, and user impersonation attack is possible in [18]. Xiong et al. proposed revocable certificateless anonymous and remote authentication scheme [13] in WBAN. But in this scheme, pairing operations do not fulfill the resource constraints. To eliminate the computational costs, non-pairing operations are utilized. Zhang et al. designed the certificateless signcryption scheme [40] to reduce the communication cost in WBAN. But the overall limitation of this scheme is the user impersonation attack and key forge ability. Shen et al. introduced the non-pairing certificateless authentication protocol [41] for WBAN using the elliptic curve. And data freshness is ensured by the checking the message authentication code values. The overall drawback of the certificateless signcryption schemes in WBAN is discussed in [42].

Attribute-Based Cryptography Lu et al. suggested a privacy-preserving authentication scheme [43] using the set of context-aware attributes in healthcare application using the access structure. Wu et al. proposed a role-based access control [8] for gaining the access of real-time data observation of the patient. But these schemes do not have resistance over key escrow problem and depends on the single authority to be trusted. In 2020, Arfaoui et al. introduced the hybrid heterogeneous signcryption scheme [44] based on both identity-based signcryption and attribute-based signcryption to eliminate the key escrow problem and user impersonation attacks. If the trust is distributed to many authorities to verify the attributes of the crypto primitives in secure channel, the high computation cost does not satisfy the resource limitations. Yao et al. suggested the ECC-based access control mechanism to eliminate the drawbacks of pairing operations, but it is not completely rectified.

Physical Unclonable Functions (PUF) In schemes like [45] and [7], the mutual authentication phase is set up through equipping the sensor nodes with PUF. Authentication takes place by sending the identity and random numbers of the sensor nodes to the sink node (local server). In this type of mechanism, the replay attack is avoided by generating different random numbers for each transaction. This is the typical example of random challenges which follows the request and response method to ensure the data freshness property.

5 Conclusion and Future Directions

In this chapter, we analyzed the feasible solutions for avoiding the replay problem while establishing the session key between the sensor node and system administrator of the medical server in WBAN. In our analysis, we identified various methods used for data freshness maintenance in WBAN. Our results are summarized as follows:

Random challenges will be highly applicable when the quality of the random numbers is improved. Although each data freshness mechanism has some drawbacks, the hybrid mechanism is well suited for avoiding the replay attack by reducing the communication and computing cost in resource-constrained environments. So these hybrid methods would be a reliable solution for attaining end-to-end data freshness. But the timestamps play an effective role in asymmetric lightweight key agreement protocols and block chain-based key agreement protocols. In this study, we have only considered the replay attacks and the related issues in WBAN. But in real time, there are various factors affecting the security and confidentiality of all the three-tier communication in WBAN. So while designing the authentication and key establishment protocol, we must be careful in terms of the cryptographic primitives, computational cost, and communication costs. In future work, data freshness in asymmetric cryptosystems such as Diffie-Hellman key agreement, bilinear mappings, ECC, PKI, and chaotic encryption of medical images might be analyzed and compared in terms of delay, end-to-end reliability, and throughput in IEEE 802.15.6.

Acknowledgments This part of this research work is supported by the Department of Science and Technology (DST), Science and Engineering Research Board (SERB), and Government of India under the TARE grant (TAR/2020/000259).

References

1. X. Meng, J. Xu, W. Liang, K.-C. Li, An anonymous mutual authentication and key agreement scheme in WBAN, in *IEEE 5th Intl Conference on Big Data Security on Cloud (BigDataSecurity), IEEE Intl Conference on High Performance and Smart Computing, (HPSC) and IEEE Intl Conference on Intelligent Data and Security (IDS)*, (2019), pp. 31–36. https://doi.org/10.1109/BigDataSecurity-HPSC-IDS.2019.00017
2. D. He, S. Zeadally, N. Kumar, J.H. Lee, Anonymous authentication for wireless body area networks with provable security. IEEE Syst. J., 1–12 (2016). https://doi.org/10.1109/JSYST.2016.2544805
3. M. Chen, S. Gonzalez, A. Vasilakos, H. Cao, V.C.M. Leung, Body Area Networks: A Survey. Mob. Netw. Appl. **16**, 171–193 (2011). https://doi.org/10.1007/s11036-010-0260-8, ID 4028196, 7 pages, 2018. https://doi.org/10.1155/2018/4028196
4. S. Kumari, M.K. Khan, M. Atiquzzaman, User authentication schemes for wireless sensor networks: A review. Ad Hoc Netw. **27**, 159194 (2015)
5. A. Gupta, M. Tripathi, T.J. Shaikh, A. Sharma, A lightweight anonymous user authentication and key establishment scheme for wearable devices. Comp. Netw. **149**, 29–42 (2019). ISSN 1389-1286
6. S. Movassaghi, M. Abolhasan, J. Lipman, D. Smith, A. Jamalipour, Wireless body area networks: A survey. IEEE Commun. Surv. Tutor. **16**(3), 1658–1686 (2014). https://doi.org/10.1109/SURV.2013.121313.00064
7. P. Gope, J. Lee, T.Q.S. Quek, Lightweight and practical anonymous authentication protocol for RFID systems using physically unclonable functions. IEEE Trans. Inform. Foren. Sec. **13**(11), 2831–2843 (2018). https://doi.org/10.1109/TIFS.2018.2832849

8. F. Wu, L.L. Xu, S. Kumari, et al., A novel and provably secure authentication and key agreement scheme with user anonymity for global mobility networks. Secur. Commun. Netw. **9**, 3527–3542 (2016)
9. Z. Mehmood, N. Nizamuddin, S.A. Ch, W. Nasar, A. Ghani, An efficient key agreement with rekeying for secured body sensor networks, in *International Conference on Digital Information Processing & Communications*, (2012), pp. 164–167
10. D. Balfanz, D.K. Smetters, P. Stewart, H. Chi Wong, Talking to strangers: Authentication in ad-hoc wireless networks, in *NDSS*, (The Internet Society, 2002)
11. X. Wu, J. Xu, W. Huang, W. Jian, A new mutual authentication and key agreement protocol in wireless body area network, in *IEEE International Conference on Smart Cloud (SmartCloud)*, vol. 2020, (2020), pp. 199–203. https://doi.org/10.1109/SmartCloud49737.2020.00045
12. R. Chen, D. Peng, Analysis and improvement of a mutual authentication scheme for wireless body area networks. J. Med. Syst. **43**(2), 1–10 (2019)
13. J. Shen, S. Chang, J. Shen, Q. Liu, X. Sun, A lightweight multi-layer authentication protocol for wireless body area networks. Fut. Gen. Comput. Syst. **78**(Part 3), 956–963., ISSN 0167-739X (2018). https://doi.org/10.1016/j.future.2016.11.033
14. E. Lara, L. Aguilar, J.A. García, Lightweight authentication protocol using self-certified public keys for wireless body area networks in health-care applications. IEEE Access **9**, 79196–79213 (2021). https://doi.org/10.1109/ACCESS.2021.3084135
15. N. Radhakrishnan, M. Karuppiah, An efficient and secure remote user mutual authentication scheme using smart cards for Telecare medical information systems. Infor. Med. Unlock. **16**, 100092. ISSN 2352-9148 (2019). https://doi.org/10.1016/j.imu.2018.02.003
16. P. Gope, T. Hwang, A realistic lightweight anonymous authentication protocol for securing real-time application data access in wireless sensor networks. IEEE Trans. Ind. Elect. **63**(11), 7124–7132 (2016). https://doi.org/10.1109/TIE.2016.2585081
17. G. Mwitende, I. Ali, N. Eltayieb, et al., Authenticated key agreement for blockchain-based WBAN. Telecommun. Syst. **74**, 347–365 (2020). https://doi.org/10.1007/s11235-020-00662-0
18. H. Xiong, Cost-effective scalable and anonymous certificateless remote authentication protocol. IEEE Trans. Informa. Foren. Sec. **9**(12), 2327–2339 (2014)
19. K.-k. Kim, M.-H. Kim, Retracted: An enhanced anonymous authentication and key exchange scheme using smartcard, in *Information Security and Cryptology-ICISC 2012. ICISC 2012*, Lecture Notes in Computer Science, ed. by T. Kwon, M. K. Lee, D. Kwon, vol. 7839, (Springer, Berlin/Heidelberg, 2012), pp. 487–494
20. M. Konan, W. Wang, A secure mutual batch authentication scheme for patient data privacy preserving in WBAN. Sensors **19**(7), 1608 (2019)
21. M. Dammak, O.R.M. Boudia, M.A. Messous, S.M. Senouci, C. Gransart, Token-based lightweight authentication to secure IoT networks, in *Proceedings of 16th IEEE Annual Consumer Communications and Networking Conference (CCNC '2019)*, (2019)
22. M. Polai, S. Mohanty, S.S. Sahoo, A lightweight mutual authentication protocol for wireless body area network, in *2019 6th International Conference on Signal Processing and Integrated Networks (SPIN)*, (2019), pp. 760–765. https://doi.org/10.1109/SPIN.2019.8711643
23. X. Li, M.H. Ibrahim, S. Kumari, A.K. Sangaiah, V. Gupta, K.-K.R. Choo, Anonymous mutual authentication and key agreement scheme for wearable sensors in wireless body area networks. Comput. Netw. **129**(P2), 429–443 (2017). https://doi.org/10.1016/j.comnet.2017.03.013
24. X. Li, M.H. Ibrahim, S. Kumari, A.K. Sangaiah, V. Gupta, K.-K.R. Choo, Anonymous mutual authentication and key agreement scheme for wearable sensors in wireless body area networks. Comp. Netw. **129**(Part 2), 429–443. ISSN 1389-1286 (2017). https://doi.org/10.1016/j.comnet.2017.03.013
25. A.M. Koya, P.P. Deepthi, Anonymous hybrid mutual authentication and key agreement scheme for wireless body area network. Computer Networks **140**, 138–151. ISSN 1389-1286 (2018). https://doi.org/10.1016/j.comnet.2018.05.006

26. A. Chaturvedi, D. Mishra, S. Jangirala, S. Mukhopadhyay, A privacy preserving biometric-based three-factor remote user authenticated key agreement scheme. J. Inform. Sec. Appl. **32**, 15–26., ISSN 2214-2126 (2017). https://doi.org/10.1016/j.jisa.2016.11.002
27. A. Sammoud, M.A. Chalouf, O. Hamdi, N. Montavont, A. Bouallègue, A new biometrics-based key establishment protocol in WBAN: Energy efficiency and security robustness analysis. Comput. Secur. **96**, 101838 (2020). https://doi.org/10.1155/2021/4939589. ID 4939589, 11 pages
28. M. Turkanović, B. Brumen, M. Hölbl, A novel user authentication and key agreement scheme for heterogeneous ad hoc wireless sensor networks based on the Internet of Things notion. Ad Hoc Netw. **20**, 96–112 (2014)
29. Q. Jiang, S. Zeadally, J. Ma, D. He, Lightweight three-factor authentication and key agreement protocol for internet-integrated wireless sensor networks. IEEE Access **5**, 3376–3392 (2017). https://doi.org/10.1109/ACCESS.2017.2673239
30. L. Xiao, D. Han, X. Meng, W. Liang, K.-C. Li, A secure framework for data sharing in private blockchain-based WBANs. IEEE Access **8**, 153956–153968 (2020). https://doi.org/10.1109/ACCESS.2020.3018119
31. D. Wang, Z. Zhang, P. Wang, J. Yan, X. Huang, Targeted online password guessing: An underestimated threat, in *Proceedings of the 2016 ACM SIGSAC Conference on Computer and Communications Security(CCS '16)*, (ACM, 2016), pp. 1242–1254
32. M. Shuai, B. Liu, N. Yu, L. Xiong, C. Wang, Efficient and privacy-preserving authentication scheme for wireless body area networks. J. Inform. Sec. Appl. **52**, 102499 (2020). https://doi.org/10.1016/j.jisa.2020.102499
33. D. He, N. Kumar, J. Chen, C.C. Lee, N. Chilamkurti, S.S. Yeo, Robust anonymous authentication protocol for health-care applications using wireless medical sensor networks. Multimed. Syst. **21**, 49–60 (2013). https://doi.org/10.1007/s00530-013-0346-9
34. P. Das, A. Vashisth, D. Chadha, S. Ananda Kumar, A. Banerjee, S. Shiaeles, LIMAP: A lightweight multilayer authentication protocol for WBAN. Wireless Per. Commun. (2021). https://doi.org/10.1007/s11277-021-08853-x
35. Z.U. Rehman, S. Altaf, S. Ahmad, S. Huda, A.M. Al-Shayea, S. Iqbal, An efficient, hybrid authentication using ECG and lightweight cryptographic scheme for WBAN. IEEE Access **9**, 133809–133819 (2021). https://doi.org/10.1109/ACCESS.2021.3115706
36. S.K. Marko Kompara, H. Islam, A. Marko Hölbl, Robust and efficient mutual authentication and key agreement scheme with untraceability for WBANs. Comp. Netw. **148**, 196–213. ISSN 1389-1286 (2019). https://doi.org/10.1016/j.comnet.2018.11.016
37. H. Lee, D. Kang, J. Ryu, D. Won, H. Kim, Y. Lee, A three-factor anonymous user authentication scheme for Internet of Things environments. J. Infor. Sec. Appl. **52**, 102494 (2020) ISSN 2214-2126
38. M. Kumar, S. Chand, A lightweight cloud-assisted identity-based anonymous authentication and key agreement protocol for secure wireless body area network. IEEE Syst. J. **15**(2), 2779–2786 (2021). https://doi.org/10.1109/JSYST.2020.2990749
39. J. Liu, Z. Zhang, X. Chen, K.S. Kwak, Certificateless remote anonymous authentication schemes for wireless body area networks. IEEE Trans. Parall. Distribut. Syst. **25**(2), 332–342 (2014)
40. Y.R. Lu, L.X. Li, H.P. Peng, et al., An enhanced biometric-based authentication scheme for telecare medicine information systems using elliptic curve cryptosystem. J. Med. Syst. **39**, 3 (2015)
41. C. Wang, Y. Zhang, New authentication scheme for wireless body area networks using the bilinear pairing. J. Med. Syst. **39**, 136 (2015)
42. P. Kasyoka, M. Kimwele, S.M. Angolo, Cryptanalysis of a Pairing-free Certificateless Signcryption scheme. ICT Express **7**(2), 200–204., ISSN 2405-9595 (2021). https://doi.org/10.1016/j.icte.2020.07.006
43. J. Hu, H.H. Chen, T.W. Hou, A hybrid public key infrastructure solution (HPKI) for HIPAA privacy/security regulations. Comput. Stand. Interf. **32**(5–6), 274–280 (2010). https://doi.org/10.1016/j.csi.2009.04.005

44. A. Arfaoui, O.R.M. Boudia, A. Kribeche, S.-M. Senouci, M. Hamdi, Context-aware access control and anonymous authentication in WBAN. Comput. Sec. Elsevier **88**, 101496 (2020). https://doi.org/10.1016/j.cose.2019.03.017. ⟨hal-02539815⟩.
45. J. Zhang, G. Qu, Physical unclonable function based key sharing via machine leaning for IoT security. IEEE Trans. Ind. Elect.. https://doi.org/10.1109/TIE.2019.2938462
46. H. Yao, Q. Yan, X. Fu, et al., ECC-based lightweight authentication and access control scheme for IoT E-healthcare. Soft Comput. (2021). https://doi.org/10.1007/s00500-021-06512-8

Vehicular Support System for User and Vehicle Accident Prevention

W. Stalin Jacob, P. Jenifer Darling Rosita, K. B. Gurumoorthy, P. Kannan, S. Allwin Devaraj, and Cynthia Anbuselvi Thangaraj

Abstract Accidents are the biggest life threat in our day-to-day life, among them accidents caused by people due to the lack of observation on dooring is being increased drastically, but with the help of electronic board digitalization in automobiles, we can design a safety door locking system to prevent happening of such accidents between vehicles, pedestrian, and passengers. In this paper, we propose a novel distance computing system using image processing; from this method the passenger cannot open their vehicle's door from inside while the moving object is in the direction of the vehicle as well as near the impact zone. If the moving obstacle is detected within the range, the relay which is connected to the central locking system of the vehicle locks the door so that we will not be able to open the door from inside; if there are no moving objects detected within the impact zone, the relay releases the door lock; hence they can open the door. We have also field tested the proposed system with various test subjects as mentioned above. Now with the help of an ultrasonic sensor and metal detector sensor, the blind spots of the vehicle can be covered, and an alert system was designed respectively, so that it will alert

W. S. Jacob
Engineering Department – Electrical Engineering, Botho University, Gaborone, Botswana
e-mail: stalin.jacob@bothouniversity.ac.bw

P. J. D. Rosita
Electrical Engineering Department, New Era College, Gaborone, Botswana
e-mail: jrosita@neweracollege.ac.bw

K. B. Gurumoorthy
Department of Electronics and Communication Engineering, KPR Institute of Engineering and Technology, Coimbatore, India
e-mail: gurukb85@gmail.com

P. Kannan · S. A. Devaraj (✉)
Electronics and Communication Engineering Department, Francis Xavier Engineering College, Tirunelveli, India

C. A. Thangaraj
SEA College of Engineering and Technology, Bangalore, India

A. Haldorai et al. (eds.), *4th EAI International Conference on Big Data Innovation for Sustainable Cognitive Computing*, EAI/Springer Innovations in Communication and Computing, https://doi.org/10.1007/978-3-031-07654-1_15

the drivers and prevent accidents that occur in the intersections on highways. The whole system is controlled with the help of an advanced microcontroller.

Keywords Dooring · Central locking system · Blind spots detection · Relay · Ultrasonic sensor

1 Introduction

In recent years, the number of cars owned by the people in India has increased rapidly. At the same time, bike-sharing has also developed rapidly in India. More and more people are choosing to ride-sharing bikes to reduce vehicle emissions for achieving environmental protection. While cars and bikes facilitate people for travel and transportation, they also cause frequent traffic accidents. Among them, the accident of colliding with the vehicles when the door of the vehicle is opened seems to be more frequent than expected; these accidents often happen between a vehicle on-road and with parked vehicles; it occurs when suddenly the door of the car was opened by the passenger as well as for the driver. These kinds of accidents pose a threat to the driver himself and to the nearby vehicle users resulting in serious injury or even death. Accidents are the biggest life threats nowadays, but with the help of electronic board digitalization in automobiles, we can make a drastic change and can make our rides safe. The concept is to make a dooring accident prevention system along with the blind spot area covering system. The reason is that there is an increase in the rate of accidents on the roads, and dooring accident is one of the serious issues, but it is always not being considered much. The theme is to use image processing to alert the user and prevent them from causing as well as getting caught in an accident. The design we proposed is user-friendly and also economically feasible.

2 Related Works

Mr. Ravindra Wakchaure, Mr. Sandip Supekar, and Mr. Kunal Tattu have proposed a smart night vision system for automobiles. The system developed mostly by adopting infrared cameras and computer vision technology aims at improving the safety and convenience of night driving. With 6175 charges left in the first 6 months, young drivers were the minimum likely to buckle up the seat belts. In the year 2011, an analysis conducted, and it was found that 55% of university students did not regularly wear a seat belt. The safety locking system is designed to prevent accidents from happening while opening the car door and to provide protection to passengers and pedestrians. To prevent accidents due to passenger's inattention, this planned to control the car door locking automatically using a distance computing system, child lock, and ultrasonic sensors. Due to this passenger cannot open their car door from the inside while the object is moving in the direction of the car. In this proposed system, we used the ultrasonic sensor to detect obstacles. If the obstacle is detected within the range, the car door will not open from the inside [1–6].

Mr. Da Li, Mr. Bodong Liang, and Mr. Weigang Zhang have proposed the method for moving vehicle detection, tracking, and counting which are very critical for traffic flow monitoring, planning, and controlling. The video-based solution, when compared to other techniques, does not disturb traffic flow, and it can be easily installed. By analyzing the traffic video sequence recorded from a video camera, this paper presents a video-based solution applied with adaptive subtracted background technology in combination with virtual detector and blob tracking technologies. Experimental results, implemented in Visual C++ code with OpenCV development kits, indicate that the proposed method can detect, track, and count moving vehicles accurately [7–10].

Mrs. N. Nagamma, Dr. M.V. Lakshmaiah, and Mrs. T. Narmada have also proposed that in biometric authentication-based vehicle door locking system, biometric authentication plays an important role to provide high security. Nowadays security is very important in order to keep our data confidential from unauthorized persons. The main aim of this paper is to protect the car from unauthorized people by using the unique ID that is fingerprint authentication. At the place of the car door locking system, the fingerprint scanner is placed to lock and unlock the doors in place of the conventional door locking system, which gives more protection to the car owner. The other indication system is also developed by using the GSM module to send the message to the car owner mobile [11–15].

3 Proposed System

Figure 1 represents the block diagram and the function of the image capturing and processing the video by separating each frame so that we can find the difference between each frame, and if the difference meets the required size with the help of contours, we can identify them and separate them as what type of object it is, and we can easily identify moving vehicles such as bikes cars and trucks so that the signal will be sent to the relay and the relay will trigger the locking system and the vehicle, and this will be intimated to the user with the help of a buzzer as well as the LCD display.

The secondary system helps us to provide a blind spot alert system with the help of an ultrasonic sensor and metal detector sensor, and for indication, we use the buzzer to give a sound alert to the driver as well as displaying the message in the LCD display that has been placed in the dashboard of the vehicle. This system is very useful for heavy-duty trucks and public transportation vehicles.

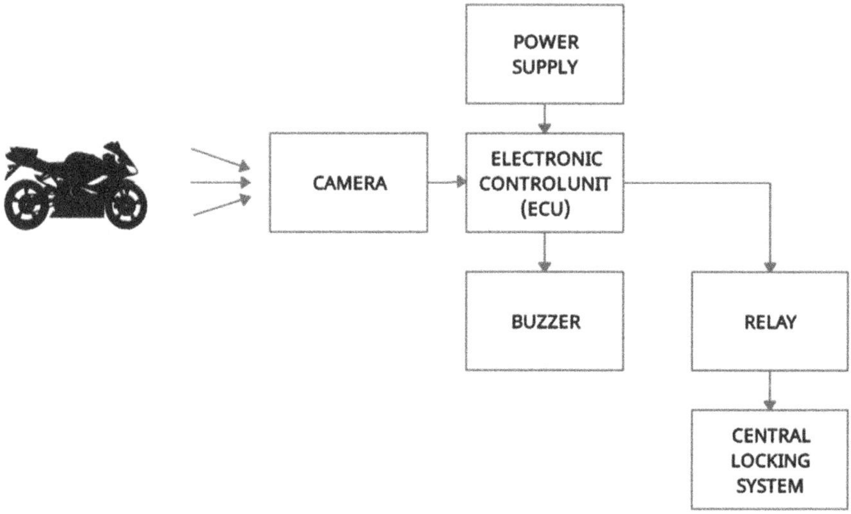

Fig. 1 ECU interfacing with central locking system

3.1 *Workflow for Framing Conversion*

Figure 2 represents the moving vehicle; according to the diagram, as a first step we keep a frame named reference frame as a none frame which is nothing but an empty black screen. Once when the camera is turned *on*, the first frame is made as a reference frame, and the next frame is considered as an actual frame, and then both the frames are converted into grayscale images, and then a Gaussian Blur is added to the images; this will help us to remove the noises from the images, and then with the help of the absolute difference between these images and then the threshold is adjusted and the image is dilated in order to remove negligible differences caused due to the noise of the images after this all the contours are detected after this process there will a condition through which we can remove the contours which have a less value. This method will remove the small differences such as tree leaves moving due to wind and shadows; it will also be helpful for not detecting the vehicle that is not moving or the pedestrian who is walking in the non-impact zone of the vehicle; after removing all these contours, we can get the desired output.

Here we set the contour's minimum value from 5000 and above; this value was obtained by a trial and error method during the field testing of the support system. After all these actions, a signal will be sent to a relay which will be connected to the central locking system, and the message "NOT SAFE TO OPEN" will be displayed in the LCD display which will be placed in the dashboard of the vehicle to indicate the drivers. After this process, the actual frame will be made as to the reference frame, and the next frame will be considered as the actual frame. All these activities will take place when parking brakes are applied, and with the help of the kill switch, we can manually disable this whole action; this switch can be

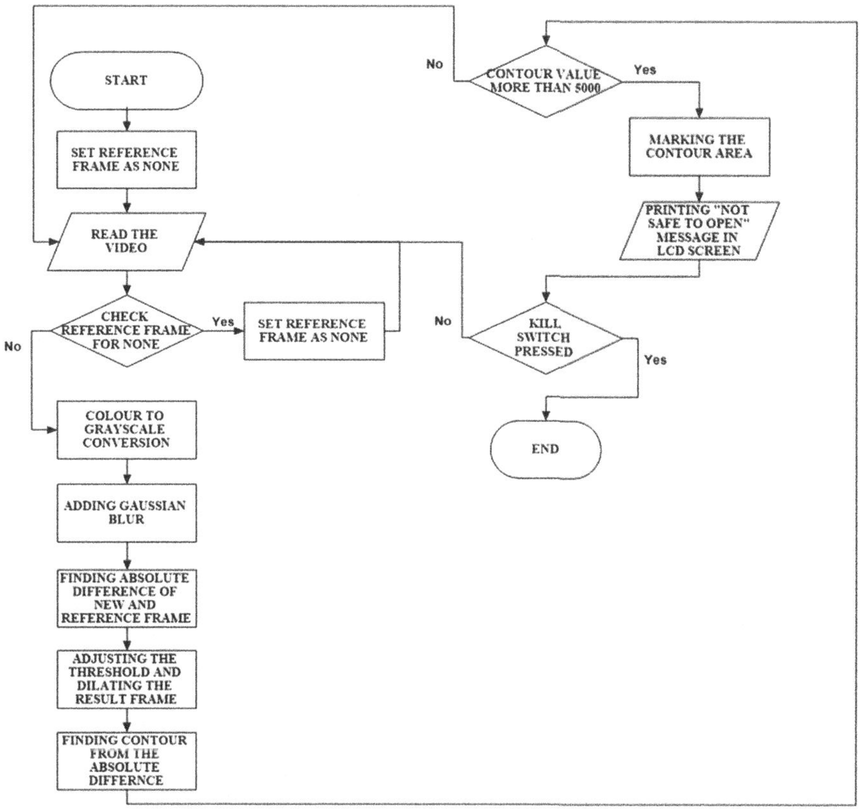

Fig. 2 Flowchart for identification of sequence

fixed in the vehicle keys which are used to lock and unlock vehicles. The design process is a very easy and simple method, and it gives us a very effective solution to prevent dooring accidents. With the help of the field test result, we found out that this system gives a very effective solution, and it detects pedestrians, bicycle, and all other vehicles with good accuracy and a very small amount of delay in the transmission of messages to the LCD display.

4 Results and Discussion

The design that has been proposed is a vehicular support system for user and vehicle prevention which can be used to prevent dooring accidents with the help of image processing and with the help of ultrasonic and metal detection sensors to detect any objects in the blind spot of the vehicle. The below images represent the experimental setup and the direct results gained from the designed system.

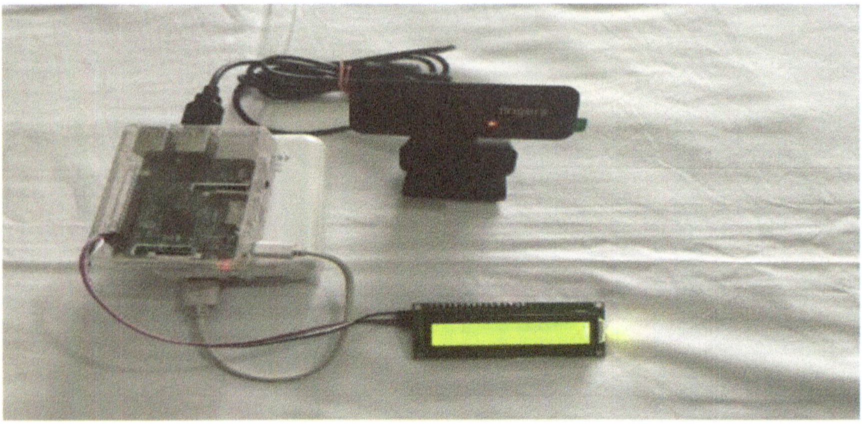

Fig. 3 Interconnection of controller with enabling device

Fig. 4 Frame structure for the various reference frame

Figure 3 shows the requirement and experimental setup for the detection of vehicles.

Figures 4 and 5 show the reference frame and the moving vehicle which is not detected. The cause for non-detection of moving vehicles is that the vehicle is far away from our desired range of field.

Figure 6 shows the pedestrian crossing our range of field near the car is getting detected by the camera.

Figure 7 shows the non-detection and detection of the vehicle. If the vehicle in the first frame is not in the range, then detection won't take place, whereas the same vehicle range is detected, and the alert message is shown to the user.

Fig. 5 Not safe zone with comparison of frames

Fig. 6 Unknown pedestrian detection with various comparison of frames

5 Summary

The propose a dooring accident prevention system based on obstacle detection using a camera and a microcontroller that act as a processing unit. The proposed method consists of three parts: frame capturing, image processing, and faster obstacle detection. When the vehicle is stopping, the camera is automatically turned on and captures the rear information of the vehicle. Then we use a simple and effective method by using OpenCV image processing features to correct the distorted images. After that, we employ the absolute difference method to compare the images and detect the moving objects, and noise is eliminated. Then we have used the LCD display to display the warning message. We also used an ultrasonic and metal detector sensor to warn the driver about the status of the blind spots of the vehicle. In future work, we aim to lock the doors automatically and by not allowing the driver

Fig. 7 Vehicle non-detection and detection

or the passengers to open the door when a moving obstacle is found to be parallelly moving close to the vehicle.

References

1. R. Wakchaure, S. Supekar, K. Tattu, Smart vehicle system for accident prevention. Int. Res. J. Eng. Technol. (IRJET) (2020)
2. M. Zhu, L. Han, F. Liang, C. Xi, L. Wu, Z. Zhang, A novel vehicle open door safety system based on cyclist detection using fisheye camera and improved deep convolutional generative adversarial nets, in *Intelligent Vehicles Symposium*, (June 2019)
3. N.N. Nagamma, M.V. Lakshmaiah, T. Narmada, Raspberry Pi based biometric authentication vehicle door locking system, in *International Conference on Power, Control, Signals and Instrumentation Engineering (ICPCSI)*, (2017)
4. P. Krishnan, Design of collision detection system for smart car using Li-Fi and ultrasonic sensor. IEEE Trans. Vehic. Technol.. 67 (12), 11420–11426. 0018-9545 (2018)
5. A. Vel'as, M. Kutaj, M. Ďurovec, Influence of changing the parameters of the camera system on video-based motion detection, in *2017 International Carnahan Conference on Security Technology (ICCST)*, (IEEE.978-1-5386-1585-0/17/, 2017)
6. B.L. Da Li, W. Zhang, Real-time moving vehicle detection, tracking, and counting system implemented with OpenCV, in *2014 4th IEEE International Conference on Information Science and Technology*, (IEEE, 978-1-4799-4808-6, 2014)

7. S.S.G. Bagi, H.G. Garakani, B. Moshiri, M. Khoshnevisan, Sensing structure for blind spot detection system in vehicles, in *International Conference on Control, Automation and Information Sciences (ICCAIS)*, (2019)
8. J. Verhaevert, Detection of vulnerable road users in blind spots through bluetooth low energy, in *Progress in Electromagnetics Research Symposium – Spring (PIERS)*, (May 2017)
9. P. Pyykonen, A. Virtanen, A. Kyytinen, Developing intelligent blind spot detection system for heavy goods vehicles, in *IEEE International Conference on Intelligent Computer Communication and Processing (ICCP)*, (IEEE. 978-1-4673-8200-7/15, 2015), pp. 293–298
10. M.K. Hossen, S.H. Tuli, A surveillance system based on motion detection and motion estimation using optical flow, in *International Conference on Informatics, Electronics and Vision (ICIEV)*, (2016)
11. R.P. Raghatate, S.S. Rajurkar, M.P. Waghmare, P.V. Ambatkar, Night vision techniques and their applications. Int. J. Mod. Eng. Res. (IJMER) **3**(2) (2013)
12. A.S. Ahire, Night vision system in BMW. Int. Rev. Appl. Eng. Res. **4**(1) (2014)
13. J. Lakshmi, C.R. Rashmi, S. Roopa, Pedestrian detection system for night vision application to avoid pedestrian vehicle related accidents. Int. J. Eng. Res. & Technol. (IJERT) **4**(07) (2015)
14. E. Schubertm, F. Meinl, M. Kunert, W. Menzel, Clustering of high-resolution automotive radar detections and subsequent feature extraction for classification of road users, in *16th International Radar Symposium (IRS)*, (Dresden, 2015), pp. 174–179
15. R.O. Chavez-Garcia, O. Aycard, Multiple sensor fusion and classification for moving object detection and tracking. IEEE Trans. Intel. Transport. Syst. **99**, 1–10 (2015)

Part V
Bigdata in Medical Applications

Application of Multi-Focused and Multimodal Image Fusion Using Guided Filter on Biomedical Images

Nalini S. Jagtap and Sudeep D. Thepade

Abstract Nowadays, the importance of image fusion in medicine has increased with the advancement of technology. Image-based medical reports are used to identify diseases or infected organs. The image quality depends on the light intensity, resolution, and focal length of the lens, cameras, etc. Over time, equipment will malfunction, reducing the quality and accuracy of images, threatening patients, and causing doctors to misdiagnose. Multimodal and multifocal models can provide clear image reports for various medical diagnosis problems. A new weighted average method based on controlled filtering is proposed to make full use of spatial consistency for baseline merging and granularity. Experimental results show that the proposed method can achieve the most advanced combination of multispectral, multifocal, and multimodality and the performance of multiple exposure images.

Keywords Guided filter · PCA · MRI · CT · SPECT · Image fusion

1 Introduction

A novel technique for combining images was developed in recent years to improve the quality of the images processed. To generate a picture suited for a particular outcome, various image sources are combined in a process called image fusion. Fusion combines several pictures for improved outcomes and user understanding compared to showing multiple images. Robotics, medical imaging, remote sensing, advances, medical diagnosis, surveillance, and forensic analysis are just a few of

N. S. Jagtap (✉)
Pimpri Chinchwad College of Engineering, Pune, Maharashtra, India

Dr D Y Patil Institute of Engineering Management and Research, Pimpri-Chinchwad, Maharashtra, India

S. D. Thepade
Pimpri Chinchwad College of Engineering, Pune, Maharashtra, India

A. Haldorai et al. (eds.), *4th EAI International Conference on Big Data Innovation for Sustainable Cognitive Computing*, EAI/Springer Innovations in Communication and Computing, https://doi.org/10.1007/978-3-031-07654-1_16

the domains where this approach is employed. Using picture fusion has shown useful in pre-diagnosis and illness categorization [1]. Use of image registration and segmentation are two common methods for composing a picture. Accordingly, combining images can provide a superior output image from a single one of its inputs. This means that image fusion's primary objective should be the creation of pictures that are valuable for a variety of applications, as well as instructive and easily classifiable images.

To see human anatomy, assess morphology and the neurological system as well as musculoskeletal restoration were created in the early 1900s with the introduction of radiography (X-ray). "Magnetic resonance spectroscopic imaging" (MRSI), "fluorescence endoscopy" (FE), "positron emission tomography" (PET), and CT are also widely employed in clinical practice, although are less well-known.

To begin with, medical image merging is concerned with the development and replication of diverse medical diagnostic pictures. Geographic information systems (GIS) for global positioning, military services, etc., employ fused pictures, weather forecasting systems, and other biomedical engineering fields. In the past 10 years, a variety of imaging technologies have been employed in medicine, as illustrated in Fig. 1.

A variety of sensing and imaging technologies have been developed as a result of technology advancements, and multimodal equipment has been offered for

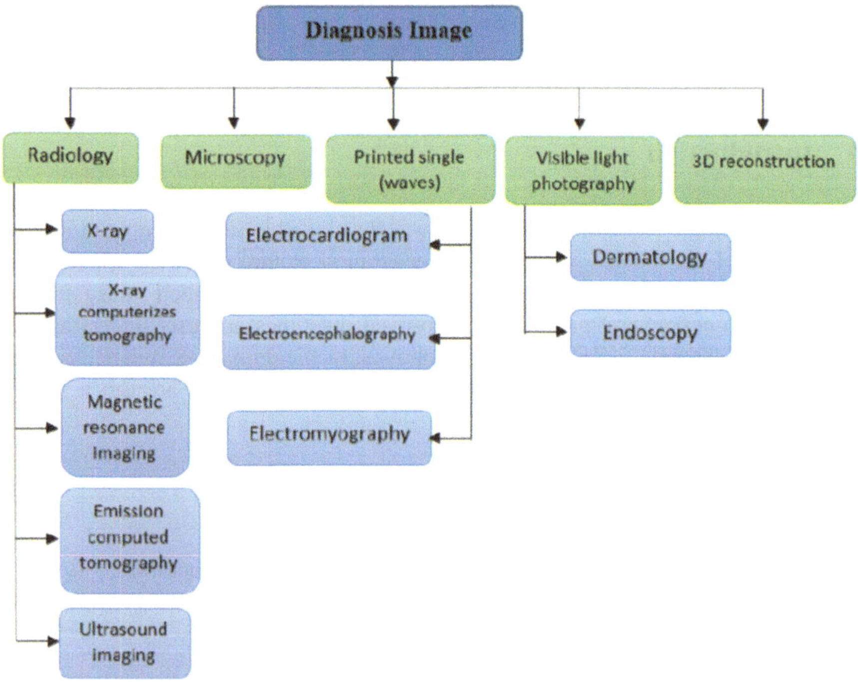

Fig. 1 Various medical imaging modalities

medical diagnosis and study. A few examples of medical imaging technologies that are underused in medical research are magnetic resonance tomography, PET, and computed tomography (CT) [2]. As a result of the fast development of computer and imaging technology in recent years, diagnostic technology has acquired a great deal of traction in the medical sector. Furthermore, these technologies have considerably simplified the diagnosis of disease and made it simpler to utilize. A typical image fusion includes merging many comparable input pictures into a single, clear fusion image by using relevant datasets. In order to assure clear and distortion-free performance, image fusion integrates and aggregates needed information from a series of recorded pictures. As a result, a large number of distinct pictures may be merged to provide a complete and undistorted set of information rich images [3]. We choose guided image filters (GF) for extracting important information from detail layers. This is an edge-preserving filter which provides the most accurate image details [4–9]. The outcome of the filtering is a linear transform of guidance image that is applied locally. It provides reduced ringing-blurring imaging effects. It is a non-iterative method with less computing. This is, to the best of our knowledge, one of the quickest edge-preserving filters.

According to the findings of the experiments, the suggested approach has properties that are equivalent to the most recent generation of picture fusion methods. The two-scale decomposition of multi-focus and multimodal pictures allows for quick fusion and filtering. The weight maps increase the details and minimize the noise in the detailed layers following decomposition. This results in more comprehensive and densely packed information in the fused pictures. The effect of pixel visibility and spatial consistency is regulated in this work by altering the filter settings. The guided filter, on the other hand, may be used for purposes other than smoothing: using the guiding picture, it can make the filtering output more structured and less smoothed than the input.

The rest of the paper is structured as follows: Section 2 gives an overview of guided image filtering process. Section 3 describes the proposed methodology in brief. Section 4 presents experiments and analysis of results. Section 5 concludes the paper.

2 Overview of Guided Filter

Smoothing is performed by guided filter by taking statistical features of a pixel's neighborhood into account. It computes output in the same way as a LTI (linear time-invariant) filter does. It does, however, utilize another picture as a guide. This extra picture might be the original or a translated version of the original. For this reason, one can alternatively use an entirely different picture. GF, like other edge-preserving filters, can maintain edge information throughout the decomposition process, which aids in the prevention of ringing artifacts. Because of this feature, GF is helpful in a variety of applications such as colorization, up-sampling, and picture matting. GF has additional attribute called structure transferring property in addition to edge-preserving property. If the guiding image and the input are the same, edge-preserving smoothing will be performed, but the structure will remain intact. The smoothing process is regulated by the structure of the guiding image when it differs from the input [10, 11].

The guided image filter, in principle, is based on the assumption that the filtering output O is a linear modification of the guiding image I in a local window ω_k centered at pixel k.

$$O_i = a_k I_i + b_k \forall_i \epsilon \omega_k \tag{1}$$

where square window is denoted by ω_k and its size is $(2r + 1)\, x\, (2r + 1)$. In ω_k linear coefficients, i.e., a_k and b_k, are the constant and can be calculated by minimizing the squared difference among output picture O and input image P.

$$E\,(a_k, b_k) = \sum_{i\epsilon\omega k} \left((a_k I_i + b_k - P_i)^2\right) + \epsilon a_k^2 \tag{2}$$

Regulation parameter is denoted by ϵ. Given below is the linear regression equation which is useful to find the linear coefficients a_k and b_k.

$$a_k = \frac{\frac{1}{|\omega|}\sum_{i\epsilon\omega_k}\left(I_i P_i - \mu_k \overline{P_k}\right)}{\delta_k + \epsilon} \tag{3}$$

$$b_k = \overline{P_k} - a_k \mu_k \tag{4}$$

Below are the parameters in the ω_k from the equation

μ_k = mean of the I.
δ_k = variance of the I
$|\omega|$= number of pixels
$\overline{P_k}$ = mean of P

According to Fig. 2, local windows are centered in the pixel, at pixel "k," whereas in the window ω_i, the pixel is "i." Hence the value of O_i will change when it is computed to another window, i.e., ω_k. This problem can be solved by taking the average of all the linear coefficients (a_k and b_k), and then filtered output is given by

$$O_i = \overline{a}_i I_i + \overline{b}_i \tag{5}$$

where

$$\overline{a}_i = \frac{1}{|\omega|}\sum_{i\epsilon\omega_k} a_k$$

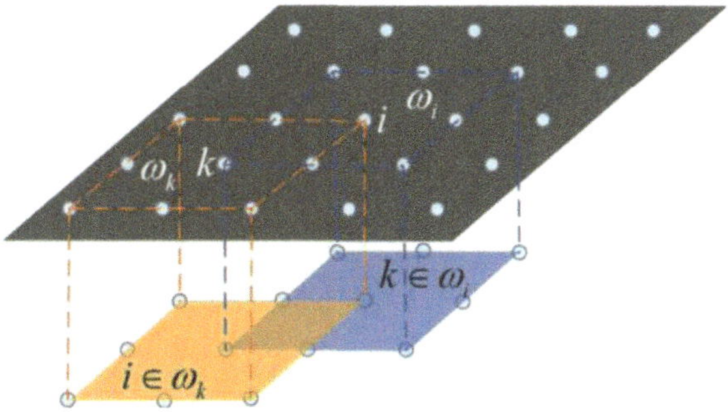

Fig. 2 Illustration of window choice

$$\overline{b_i} = \frac{1}{|\omega|}\sum\nolimits_{i\epsilon\omega_k} b_k$$

For a guided filtering, operation is denoted by the $\mathrm{GIF}_{r,\epsilon}(P, I)$ with filter size decided by r and ϵ decides the blur degree of GF. Furthermore, P is the input picture, while I denotes the guidance image.

If the input is a colored picture, the output will then be produced by applying the guided filter to the input channels, i.e., red, green and blue, respectively. If guidance image I is a colored image, then guided filter will be enlarged by the steps as shown below. The output equation can be rewritten as follows:

$$O_i = a_k^T I_i + b_k \forall_{i\epsilon\omega_k} \tag{6}$$

where a_k and I_i are the coefficient vector and colour vector of the size 3×1, respectively. Similarly, Eqs. (3), (4), and (5) can be rewritten as

$$a_k = \left(\sum\nolimits_k + \epsilon U\right)\left(\frac{1}{|\omega|}\sum\nolimits_{i\epsilon\omega_k} I_i P_i - \mu_k \overline{P_k}\right) \tag{7}$$

$$b_k = \overline{P_k} - a_k^T \mu_k \tag{8}$$

$$O_i = \overline{a}_k^T I_i + \overline{b}_i \tag{9}$$

where $U =$ Identity matrix and $\sum_k =$ Covariance Matrix with matrix size of 3×3.

Figure 3 depicts two instances of the guided filtering process. In pictures 1 and 2, (a) and (c) represent the input images, while (b) and (d) represent the output images. Image (b) is the filtered image with ϵ =0.3 and $r = 15$, where images (a) and (c) serve as the guidance and input images, respectively. Similarly, image (d) is

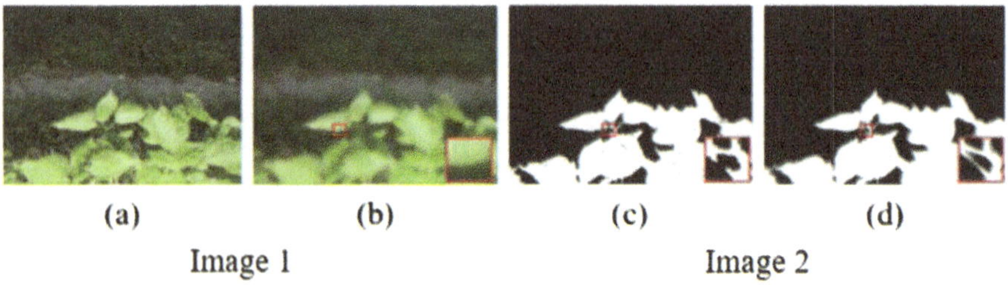

Fig. 3 In this figure, (a) and (c) represent a color guide

a filtered image with $r = 10$ and $\epsilon =,10^{-7}$, with images (a) and (c) serving as the guidance and input images, respectively.

In this example, Image 1 (a) in Fig. 3 depicts the color image with the size "620x464". To create the filtered picture, a guided image filter is applied to each of the image (a)'s color channels. In the figure 3(a) and (c) represent a color guide picture and an input image, respectively. Figure 3(b) depicts a close-up view image in which the image's features are softened, yet the image's strong edges are preserved. The input picture in Fig. 3c is noisy and poorly aligned with the object borders. Following guided filtering, as shown in Fig. 3d, noisy pixels are eliminated, and the filtered image's edges are aligned with object boundaries. It demonstrates that pixels with similar colors in the guided picture are likely to have similar range in the filtering process.

Another example, illustrated in Fig. 4, shows the effect of the window radius and regularization parameter on the base and detailed layers of the input pictures. The picture below depicts the outputs of the base and detailed layers for various r and ε values. When the variable $\varepsilon = 0.1^2$, the smoothed subband pictures have a strong edge enhancement effect, but the detailed subband images have fewer notable characteristics; when the variable $= 0.4^2$. Smooth information is spread across the detailed subband pictures because the smoothness is too high. Hence, by controlling these variables, more accurate fused images can be extracted from the filtering process.

3 Methodology

The proposed research aims to improve processing quality through picture fusion utilizing two-scale image decomposition and the guided filtering approach. The following are the stages for image fusion using a guided filter:

- The pictures in the input dataset are decomposed into detailed layers and base layers in the first step.
- The average Z filter is used to extract the base layers from the source pictures.

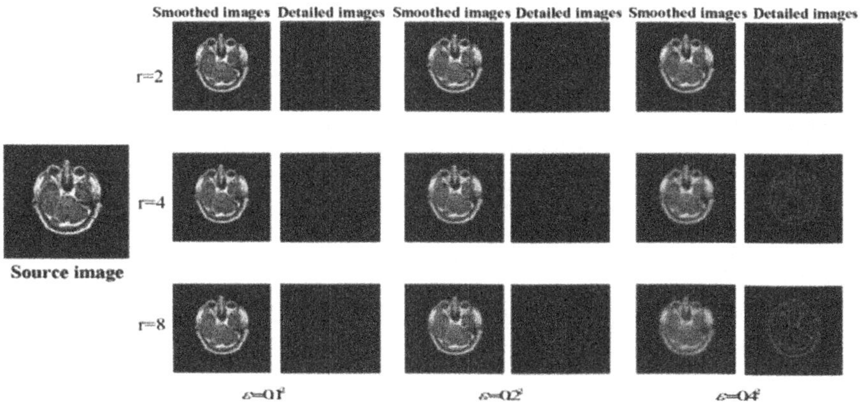

Fig. 4 Guided filtering results by varying radius and regularization parameters [12]

- The detailed layers are then calculated by subtracting the base layers from the source pictures.
- A simple averaging method is employed to fuse the base layers.
- For fusing all the detail layers:

  a) The Laplacian filter is used to create high-pass pictures.
  b) Saliency maps are created using Gaussian filters on H pictures.
  c) To generate initial weight maps, the saliency maps are contrasted.
  d) The acquired maps are then put through the guided filter. To generate the final weight maps, the matching detail layer images are used as a reference image.
  e) The final detailed layer weight maps are fused using the pixel-wise weighted averaging approach.
  f) Finally, to recreate the resultant fused image, the fused base layer and detailed layers map are utilized.

Two-Scale Decomposition

For each of the source images, a two-scale decomposition is conducted to create a base layer that captures complex structural details (information) and a detail layer that contains texture details and noises. For a pair of MRI and CT, as illustrated in Fig. 5, let $\mathrm{GIF}_{r,\varepsilon}(i,j)$ describe the action of guided image filtering, where r = radius *of* local window and ε = regularization factor of guided image filter, respectively, and (i,j) represent the input and guidance images, respectively.

(a) **Base Layer Fusion**

The input image is firstly decomposed into two-scale base layers of $B_1$and B_2 image representations by average filter Z,and the filter size is usually is set to $31 * 31$ [13].

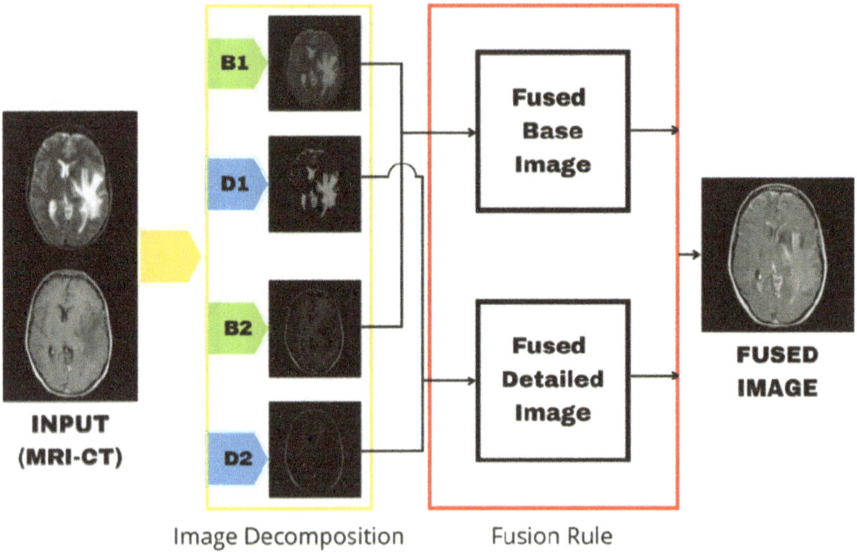

Fig. 5 Schematic diagram of the proposed framework

$$B_1 = A * Z \quad (10)$$

$$B_2 = B * Z \quad (11)$$

The fusion rule for the base layer images approximates the original images. Therefore, simple averaging scheme can be used for fusing the base layers.

$$B_f = (B_1 + B_2) * 0.5 \quad (12)$$

(b) **Detailed Layer Fusion**

The detailed layers hold the remaining of the texture features and noises. A weighted average approach based on guided filtering is used for detail layer fusion, which increases the contrast level as much as feasible through noise filtering. In general, because the Laplacian operator detects edges by conducting differentiation operations on pictures, it is sensitive to discrete points or disturbances. As a result, Laplacian filtering is applied to pictures, followed by Gaussian convolution filtering to lessen noise and improve the resilience of the image to noise. Secondly, saliency map is generally built with the following key elements in mind: detection accuracy and computational complexity. The Gaussian function, on the other hand, is simply controlled by a few parameters like blurring degree and filter size in order to create a more comprehensive saliency map. The saliency maps of detailed layers are created using Laplacian filtering, and later the Gaussian filter is applied for smoothing to highlight small-scale edge characteristics and denoise the detail layers. The following is a thorough description of the implementation [14].

To produce the detail layers D_1 *and* D_2, the absolute values of the variance between the source and base layers are calculated.

$$D_1 = \mid A - B_1 \mid \tag{13}$$

$$D_2 = \mid B - B_2 \mid \tag{14}$$

where $\mid \cdot \mid$ = absolute value operator. Weight maps are constructed using Laplacian Filters. The very first step to get the high-pass image H by applying the Laplacian filtering to D_1 and D_2 expressed as

$$H_1 = D_1 * L_{3\times 3} \tag{15}$$

$$H_2 = D_2 * L_{3\times 3} \tag{16}$$

where $L_{3\times 3}$ = Laplacian filter of 3×3 size. For saliency map construction, a Gaussian filter is employed to give local smoothing to the H-maps. The saliency maps are given as

$$S_1 = |H_1| * G_{\alpha,\sigma} \tag{17}$$

$$S_2 = |H_2| * G_{\alpha,\sigma} \tag{18}$$

where $G_{\alpha,\sigma}$=Gaussian filter of size $(2\alpha + 1) \times (2\alpha + 1)$, σ= blurred degree. To construct the first weight map, the saliency maps are compared.

$$w_1(x, y) = \begin{cases} 1, S_1(x, y) > S_1(x, y) \\ 0, \text{Otherwise} \end{cases} \tag{19}$$

$$w_2(x, y) = 1 - w_1(x, y) \tag{20}$$

where these initially generated weight maps are a binary map corresponding to "choose max" approach in the pixel location (x, y). Each initial weight map is then subjected to guided filtering, and the associated detail layer serves as the guide picture for improving the initial weight map.

$$W_{f1} = \text{GIF}_{r_2\varepsilon_2}(w_1, D_1) \tag{21}$$

$$W_{f2} = \text{GIF}_{r_2\varepsilon_2}(w_2, D_2) \tag{22}$$

where W_{f1} and W_{f2} are the final weight maps of D_1 and D_2, respectively. In this step, to retrieve the remaining edge information and establish spatial consistency, a smaller window radius and regularisation factor are utilized. To generate the fused detail layer, pixel-wise weighted averaging method is used:

$$D_f = W_{f1}D_1 + W_{f2}D_2 \tag{23}$$

Image Reconstruction
Finally, we combine the fused base layer B_f with the fused detail layer D_f to get the final fused image.

$$I_f = B_f + D_f \tag{24}$$

4 Experimental Analysis

4.1 Image Quality Assessment

There is a significant degree of structure in natural picture signals: their pixels display strong relationships, especially when they are spatially close, and these dependencies contain crucial information about object structure. Most of the quality metrics are based on linear transformations to decompose the picture signals. To prove the effectiveness of any fusion process, the findings must be verified. In terms of visual quality and fusion parameters, it's possible to accomplish this. In every image processing scenario, visual examination is the best approach to assess the quality of the final image. As a way to help the validation process, quantitative analysis will be explored. To date, there is no benchmark for calculating the proper fusion metrics in picture fusion for quantitative analysis. Traditional and contemporary fusion measures are discussed in this work. The table gives an overview on different fusion metrics used in this paper.

4.2 Subjective and Objective Analysis

We present several experimental validations of the proposed approach in this section. The input dataset contains CT, PET, MRI, and SPECT images of the human brain. Subjective and objective evaluations are used to test and validate the suggested method. Simultaneously, the proposed image fusion method is pitted against seven different standard image fusion algorithms. These conventional algorithms are PCA (principal component analysis), NSCT (nonsubsampled contourlet transform), DWT (discrete wavelet transform), GIF (graphics interchange format), CNN (convolution neural network), PCNN (pulse-coupled neural network), and NSCT-GIF. Tables 1 and 2 present a thorough study of performance for traditional and theoretical hybrid fusion methods for quality measures such as mSSIM, IQI, and PSNR

Tables 2 and 3 are also represented graphically for gaining a clear picture about the efficiency of the proposed guided filter approach. Figure 6a, b, c depicts the parameter comparative analysis of different methods with the proposed fusion method.

Figures (a), (b), and (c) exhibit bar graphs that further demonstrate the performance level of the suggested hybrid approach in comparison to traditional methods. In the image quality index Graph (b), the proposed system outperforms methods such as PCA, GIF, CNN, DWT, NSCT, PCNN, and NSCT-GIF, with practical simulated values of 0.99, which is extremely near to 1. Similarly, mSSIM is utilized to calculate mean structural similarity in Graph (a). This parameter evaluates structural similarity, as described in the parameter table. With an output of 0.99, the suggested hybrid method achieves mSSIM accuracy (almost 1). A higher PSNR indicates a higher quality restoration. The PSNR values of the prior seven techniques are in the 30–40 dB range in Graph (c); however the proposed system provides value above the 40 dB range. This means that the reconstructed fused image has a higher resolution and quality.

Many additional critical characteristics are determined to support the claims made by the proposed approach. Tables 1 and 4 describe different quality metrics parameters and output values obtained using the proposed algorithm. Sets (1, 4, 5) have positive SNR values, indicating that the image has no background noise. MSE and MAE are error numbers that indicate whether or not the pictures are identical, as well as inaccuracy in the recreated image. The developed method produces near-zero values, suggesting that the produced images are similar and accurate. The presence of reduced distortion is indicated by RMSE values close to zero. NIQE of sets (2, 3, 6) indicates that the distances between the source and fused image are less, resulting in less distortion and a better resolution. The fact that CC attained values close to one demonstrates that the images covary

Figure 7 depicts the input dataset of several imaging modalities, including MRI-SPECT, MRI-PET, and CT-MRI, which are fused to provide a clear output picture. The guided filter approach, as described above in the methodology section, is utilized to fuse these pictures. Figure 8 depicts the system's outputs. And, based on Figs. 7 and 8, we can conclude that the proposed system presents a comprehensive and detailed image.

Table 1 Function and mathematical formulae for different quality metrics

Parameter	Formula	Function
Peak signal-to-noise ratio	$\mathrm{PSNR} = 20\log_{10}(\mathrm{MAX}_i) - 10\log_{10}(\mathrm{MSE})$	Estimates the reconstruction quality of the image
Mean square error	$\mathrm{MSE} = \frac{1}{M*N}\sum_{i=0,j=0}^{M-1,N-1}[I\ (i,\ j) - K\ (i,\ j)]^2$ M, N=width and height $I(i,j)$= enhanced image $K(i,j)$= source image i, j= row and column of source and enhanced images	Determines whether or not two pictures are identical. The "0" value error indicates that the pictures are entirely similar
Signal-to-noise ratio	$\mathrm{SNR} = \frac{P_{\mathrm{signal}}}{P_{\mathrm{noise}}}$	The noise is calculated. Positive values imply that there is less background noise in pictures
Image quality index	$\mathrm{IQI} = 1 - SI$ SI= similarity index	Estimates the quality of the reconstructed image and lies in between $[-1, 1]$
Naturalness image quality estimator	$\mathrm{NIQE} = \sqrt{\left((v_1 - v_2)^T \left(\frac{\sum_1 + \sum_2}{2}\right)^{-1} (v_1 - v_2)\right)}$ $v_1, v_2, \sum_1, \sum_2$= mean vectors and covariance matrices of the original and distorted images	The distance between the deformed picture and the original image is calculated. This distance is used to assess image quality. A shorter distance implies that the degree of distortion is low and the image quality is high
Mean absolute error	$\mathrm{MAE} = \lvert E(x) - E(y)\rvert$ $E(x), E(y)$ = intensities of source and enhanced image, respectively.	Used to compute the absolute inaccuracy between the original picture and the improved image
RMSE	$\mathrm{RMSE} = \sqrt{\mathrm{MSE}}$	The RMSE can be used to assess image quality and the degree of distortion present
Pearson correlation coefficient or CC (r)	$r = \frac{\sum_{i=1}^{n}(i_1 - i)(j_1 - j)}{\sqrt{\sum_{i=1}^{n}(i_1 - i)^2}\sqrt{\sum_{i=1}^{n}(j_1 - j)^2}}$ i_1, j_1= intensity of *ith and jth* image i, j= mean intensity	Controls the degree to which the input and output pictures covary. The correlation coefficient is frequently used in statistical analysis, pattern recognition, and image processing to compare pictures

Table 2 Quality metrics comparative analysis for different fusion methods (set1, set2 and set3)

Fused images	*Set*	*Methods*	*IQI*	*mSSIM*	*PSNR*
MRI-SPECTMRI-PETCT-MRI	Set 1	PCA [15]	0.429	0.423	35.38
		DWT [16]	*0.441*	*0.438*	*37.08*
		GIF [17]	0.553	0.522	39.64
		PCNN [18]	*0.586*	*0.55*	*38.3*
		NSCT [19]	0.603	0.593	34.69
		CNN [20]	*0.908*	*0.89*	*34.7*
		NSCT-GIF [21]	0.941	0.824	30.06
		Proposed system(guided filter)	*0.99728*	*0.9979*	*42.6164*
MRI-SPECTMRI-PETCT-MRI	Set 2	PCA [15]	0.462	0.423	29.05
		DWT [16]	*0.558*	*0.548*	*32.39*
		GIF [17]	0.542	0.595	35.55
		PCNN [18]	*0.528*	*0.575*	*37.4*
		NSCT [19]	0.701	0.67	39.58
		CNN [20]	*0.92*	*0.9392*	*39.72*
		NSCT-GIF [21]	0.94	0.9412	36.62
		Proposed system(guided filter)	*0.99131*	*0.9954*	*42.9871*
MRI-SPECTMRI-PETCT-MRI	Set 3	PCA [15]	0.487	0.451	32.68
		DWT [16]	*0.521*	*0.558*	*33.76*
		GIF [17]	0.594	0.581	38.56
		PCNN [18]	*0.625*	*0.591*	*35.8*
		NSCT [19]	0.682	0.693	39.67
		CNN [20]	*0.968*	*0.9364*	*33.65*
		NSCT-GIF [21]	0.997	0.9852	45.68
		Proposed system(guided filter)	*0.99612*	*0.9972*	*45.786*

Table 3 Quality metrics comparative analysis for different fusion methods (set4, set5, and set6)

Fused images	Set	Methods	IQI	mSSIM	PSNR
MRI-SPECTMRI-PETCT-MRI	Set 4	PCA [15]	0.433	0.436	28.6
		DWT [16]	0.529	0.472	29.45
		GIF [17]	0.611	0.548	32.56
		PCNN [18]	0.652	0.589	35.32
		NSCT [19]	0.802	0.645	39.78
		CNN [20]	0.988	0.9295	23.8
		NSCT-GIF [21]	0.972	0.9545	40.67
		Proposed system(guided filter)	0.99654	0.9978	44.06914
MRI-SPECTMRI-PETCT-MRI	Set 5	PCA [15]	0.414	0.45	34.78
		DWT [16]	0.528	0.52	36.4
		GIF [17]	0.574	0.602	43.01
		PCNN [18]	0.602	0.646	40.23
		NSCT [19]	0.664	0.696	40.3
		CNN [20]	0.893	0.8505	40.2
		NSCT-GIF [21]	0.932	0.9056	32.3
		Proposed system(guided filter)	0.99662	0.9976	42.86026
MRI-SPECTMRI-PETCT-MRI	Set 6	PCA [15]	0.567	0.529	30.56
		DWT [16]	0.573	0.548	32.58
		GIF [17]	0.619	0.629	34.56
		PCNN [18]	0.596	0.605	33.3
		NSCT [19]	0.769	0.695	38.4
		CNN [20]	0.902	0.8984	37.46
		NSCT-GIF [21]	0.926	0.9804	38.4
		Proposed system(guided filter)	0.99425	0.9971	43.2782

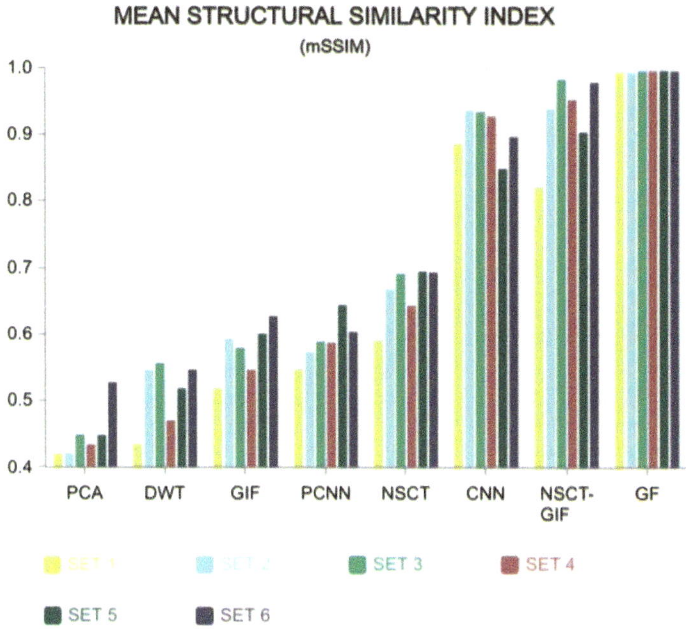

(a)

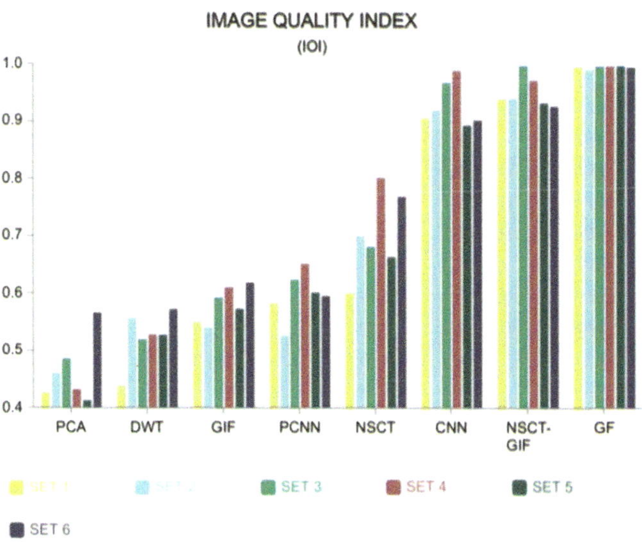

(b)

Fig. 6 Comparative analysis of different image quality metrics (mSSIM(**a**), IQI(**b**), PSNR(**c**))

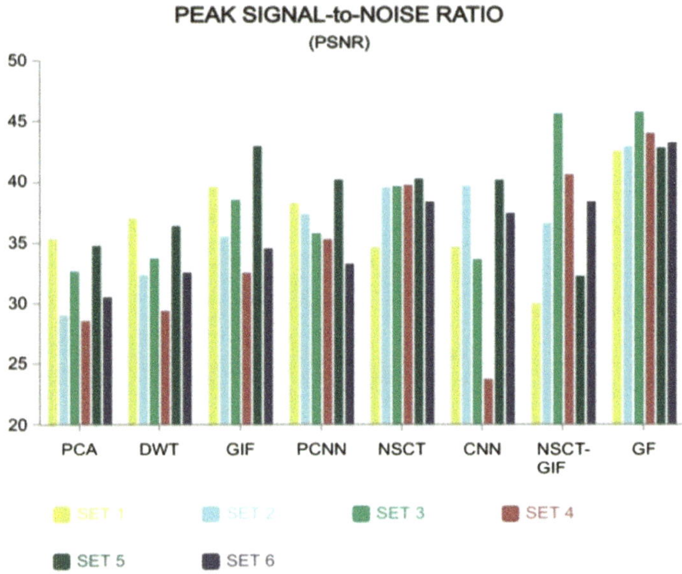

(c)

Fig. 6 (continued)

Table 4 Other parameters calculated by the proposed guided filter method

Parameters	Set1	Set2	Set3	Set4	Set5	Set6
SNR	*0.00018 dB*	*−0.00026 dB*	*−0.00007 dB*	*0.00017 dB*	*0.00019 dB*	*−0.00009 dB*
MAE	*0.0540*	*0.0426*	*0.0268*	*0.0391*	*0.0509*	*0.0484*
MSE	*0.1076*	*0.0988*	*0.0518*	*0.0770*	*0.1017*	*0.0922*
RMSE	*0.3280*	*0.3143*	*0.2277*	*0.2775*	*0.319*	*0.3040*
Niqel	*5.2601*	*3.9176*	*4.494*	*12.1482*	*22.0537*	*2.3514*
Entropy	*6.5836*	*4.847*	*4.4893*	*5.954*	*6.8825*	*4.6216*
CC	*0.9998*	*0.9998*	*0.9998*	*0.9998*	*0.9998*	*0.9999*

5 Conclusion

In this paper, we presented a multi-focus and multimodal picture fusion method based on guided image filtering and two-scale decomposition. To produce base (coarse) layers, we used basic averaging. Guided filter and saliency maps are used to reduce noise from the detailed layers. The original image's information can be rapidly and efficiently retrieved using two-scale decomposition, while a guided filter can suppress artifacts to reduce ringing and blurring effects while retaining edges and enhancing image clarity. The proposed approach was found to be valid and

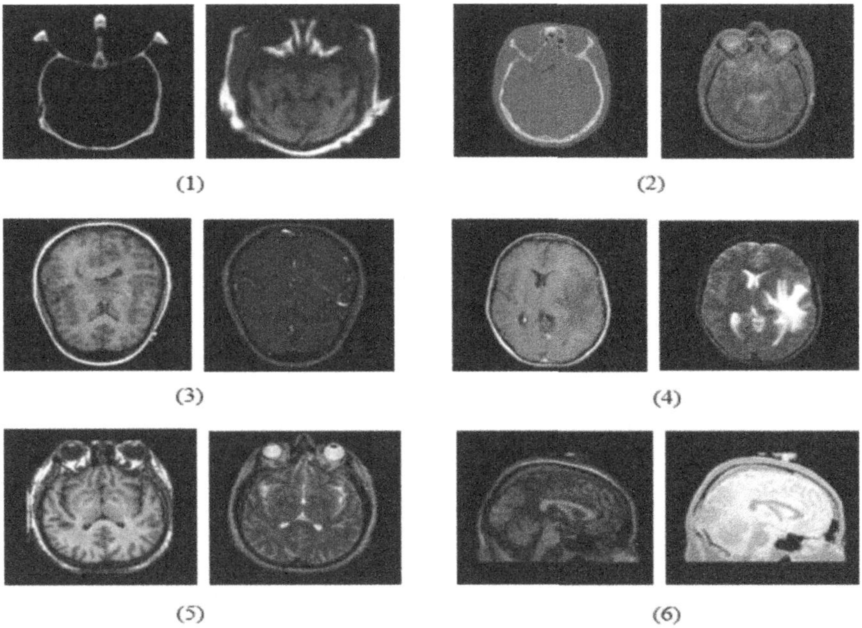

Fig. 7 Input dataset to the proposed system

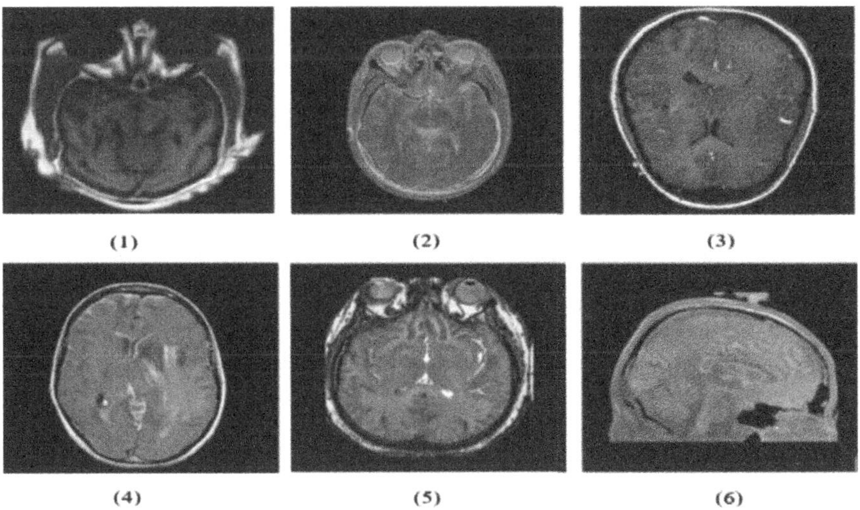

Fig. 8 Output of the system integrated with the guided filter

effective as a result of the experiments, and it performs adequately in both visual impression and quantitative performance.

References

1. S.P. Yadav, S. Yadav, Image fusion using hybrid methods in multimodality medical images. Med. Biol. Eng. Comput. **58**(4), 669–687 (2020). https://doi.org/10.1007/s11517-020-02136-6
2. R. Shen, I. Cheng, A. Basu, Cross-scale coefficient selection for volumetric medical image fusion. I.E.E.E. Trans. Biomed. Eng. **60**(4), 1069–1079 (2013). https://doi.org/10.1109/TBME.2012.2211017
3. M.G. Reddy, P.V.N. Reddy, P.R. Reddy, Medical image fusion using integrated guided nonlinear anisotropic filtering with image statistics. Int. J. Intell. Eng. Syst. **13**(1), 25–34 (2020). https://doi.org/10.22266/ijies2020.0229.03
4. D.P. Bavirisetti, V. Kollu, X. Gang, R. Dhuli, Fusion of MRI and CT images using guided image filter and image statistics. Int. J. Imaging Syst. Technol. **27**(3), 227–237 (2017). https://doi.org/10.1002/ima.22228
5. X. Wang, R. Nie, X. Guo, Two-scale image fusion of visible and infrared images using guided filter. ACM Int. Conf. Proc. Ser., 217–221 (2018). https://doi.org/10.1145/3208854.3208881
6. X. Yan, H. Qin, J. Li, H. Zhou, T. Yang, Multi-focus image fusion using a guided-filter-based difference image. Appl. Opt. **55**(9), 2230 (2016). https://doi.org/10.1364/ao.55.002230
7. T.O. Sunitha, R. Rajalakshmi, Multi-modal image fusion technique for enhancing image quality with multi-scale decomposition algorithm. Comput. Methods Biomech. Biomed. Eng. Imaging Vis. **9**(2), 192–204 (2021). https://doi.org/10.1080/21681163.2020.1830437
8. H. Singh, V. Kumar, S. Bhooshan, A novel approach for detail-enhanced exposure fusion using guided filter. Sci. World J. **2014**, 1–8 (2014). https://doi.org/10.1155/2014/659217
9. C. Rong, Y. Jia, Y. Yang, Y. Zhu, Y. Wang, X. Ni, Multi-modal image fusion via a novel multi-scale edge-preserving decomposition. 2019 11th Int. Conf. Wirel. Commun. Signal Process. WCSP 2019 (3) (2019). https://doi.org/10.1109/WCSP.2019.8927992
10. I. Journal, E. Trends, Image fusion with edge preserving smoothing filter. Int. J. Emerg. Trends Eng. Res. **3**(6), 216–222 (2015)
11. K. He, J. Sun, X. Tang, Guided image filtering. IEEE Trans. Pattern Anal. Mach. Intell. **35**(6), 1397–1409 (2013). https://doi.org/10.1109/TPAMI.2012.213
12. W. Li, L. Jia, J. Du, Multi-modal sensor medical image fusion based on multiple salient features with guided image filter. IEEE Access **7**, 173019–173033 (2019). https://doi.org/10.1109/ACCESS.2019.2953786
13. J. Du, W. Li, Two-scale image decomposition based image fusion using structure tensor. Int. J. Imaging Syst. Technol. **30**(2), 271–284 (2020). https://doi.org/10.1002/ima.22367
14. D.P. Bavirisetti, G. Xiao, J. Zhao, R. Dhuli, G. Liu, Multi-scale guided image and video fusion: A fast and efficient approach. Circuits, Syst. Signal Process. **38**(12), 5576–5605 (2019). https://doi.org/10.1007/s00034-019-01131-z
15. H. Wei et al., Free-breathing diffusion tensor imaging and tractography of the human heart in healthy volunteers using wavelet-based image fusion. IEEE Trans. Med. Imaging **34**(1), 306–316 (2015). https://doi.org/10.1109/TMI.2014.2356792
16. M.M.I. Ch, M.M. Riaz, N. Iltaf, A. Ghafoor, S.S. Ali, A multifocus image fusion using highlevel DWT components and guided filter. Multimed. Tools Appl. **79**(19–20), 12817–12828 (2020). https://doi.org/10.1007/s11042-020-08661-8
17. B. Rajalingam, R. Priya, A Novel approach for Multimodal Medical Image Fusion using Hybrid Fusion Algorithms for Disease Analysis. Inte. J. Pure Appl. Math. **117**(15), 599–619 (2017)., [Online]. Available: http://www.ijpam.eu
18. B. Huang, F. Yang, M. Yin, X. Mo, C. Zhong, A review of multimodal medical image fusion techniques. Comput. Math. Methods Med. **2020** (2020). https://doi.org/10.1155/2020/8279342
19. V. Bhateja, H. Patel, A. Krishn, A. Sahu, A. Lay-Ekuakille, Multimodal medical image sensor fusion framework using cascade of wavelet and contourlet transform domains. IEEE Sensors J. **15**(12), 6783–6790 (2015). https://doi.org/10.1109/JSEN.2015.2465935

20. Y. Liu, X. Chen, H. Peng, Z. Wang, Multi-focus image fusion with a deep convolutional neural network. Inf. Fusion **36**, 191–207 (2017). https://doi.org/10.1016/j.inffus.2016.12.001
21. B. Rajalingam, F. Al-Turjman, R. Santhoshkumar, M. Rajesh, Intelligent multimodal medical image fusion with deep guided filtering. Multimed. Syst (0123456789) (2020). https://doi.org/10.1007/s00530-020-00706-0

Modelling Cognitive Scores for Alzheimer's Disease Progression Prediction Using Longitudinal MRI Data

A. Thushara, C. UshaDevi Amma, Ansamma John, and Lida Mary Victor

Abstract Alzheimer's disease (AD) is a progressive neurological illness that causes memory and cognitive impairment. Many research works are going on to diagnose the disease at an early stage as there are no supplements or medications to decrease the risk of AD. Cognitive tests such as Mini-Mental State Examination (MMSE) and Alzheimer's Disease Assessment Scale-Cognitive Subscale (ADAS-Cog) are used to measure the level and progression of cognitive impairment, as well as to track an individual's cognitive changes over time, making it a useful tool for documenting a patient's reaction to treatment. These examinations are biased against people with hearing or vision problems and those with a poor level of education due to linguistic and mathematical examination components. The proposed study uses regression models such as ridge, lasso and elastic net to estimate cognitive scores using longitudinal MRI data from Alzheimer's Disease Neuroimaging Initiative (ADNI). Compared to ridge and lasso models, elastic net model achieves promising performance in cognitive score prediction and helps to identify the important biomarkers for AD.

Keywords Alzheimer's disease · Ridge model · Lasso model · Elastic net model

1 Introduction

Alzheimer's disease (AD) is a neurodegenerative disease that begins slowly and progressively worsens. As the disease progresses, symptoms include disorientation, communication, mood swings, self-neglect, behavioural issues and loss of motiva-

A. Thushara (✉) · A. John · L. M. Victor
Department of Computer Science and Engineering, TKM College of Engineering, APJ Abdul Kalam Technological University, Thiruvananthapuram, Kerala, India

C. UshaDevi Amma
Department of Electronics and Communications Engineering, Amrita Vishwa Vidyapeetham, Vallikavu, Kerala, India

A. Haldorai et al. (eds.), *4th EAI International Conference on Big Data Innovation for Sustainable Cognitive Computing*, EAI/Springer Innovations in Communication and Computing, https://doi.org/10.1007/978-3-031-07654-1_17

tion. The rate of progression can vary from person to person, but life expectancy after diagnosis usually ranges from 3 to 9 years. There is no treatment available to cure this disease. But the early diagnosis can slow down the rate of progression and enhance the quality of life [1]. Therefore, a lot of research works are going on to diagnose the disease at an early stage.

Based on the progressive pattern of functional and cognitive impairment, the subject under this study can be categorized as Alzheimer's disease (AD), early mild cognitive impairment (EMCI), late mild cognitive impairment (LMCI), and control normal (CN). The disease damages the memory-related brain regions, which triggers the first signs of memory impairment [2]. The brain regions such as medial temporal lobe, especially the entorhinal cortex and the hippocampus, develop substantial atrophy during the early stages of AD that affects the cognitive skills [3]. Many cognitive tests with mathematical and language components are developed to measure the cognitive score, which indicates the severity of AD [4].

These cognitive scores are highly influenced with educational attainment of subjects. Furthermore, verbal responses, reading and writing play a vital role in generating the test's cognitive score [5]. As a result, it is difficult to determine whether a low-score person is an Alzheimer's patient or someone who has hearing or vision impairment. Similarly, AD patients who repeat the same exam and achieve a higher score in a follow-up session may be misinterpreted as control normal patients. This possibility reduces the scope of the test as a tool for diagnostic purposes. Furthermore, the human expert with prior knowledge of the subject's cognitive ability may lead to biased evaluation. All these limitations associated with the cognitive test can be addressed to a certain extent by utilizing features in neuroimages captured by neuroimaging techniques which have proven to be an effective method for determining the neurodegenerative process in Alzheimer's disease. Early AD can be diagnosed by measuring the important parameters of various brain regions, which predicts the cognitive scores more accurately by non-invasive neuroimaging techniques like MRI (magnetic resonance imaging) [6]. Hence, neuroimaging-based longitudinal studies have been extensively carried out in recent decades to predict cognitive scores by utilizing data acquired at specified time intervals. Out of many AD prediction studies available, regression models has become popular to investigate the association between cognitive scores such as MMSE [7] and ADAS13 [8] and structural changes in magnetic resonance (MR) images.

The proposed work aims at predicting cognitive scores accurately from the longitudinal brain MR images by identifying and extracting the relevant features that aid in the AD diagnosis rather than conducting cognitive tests. This work also investigates the identification of the important brain regions that are affected by AD, by evaluating the correlation between the features extracted from MRI and cognitive scores. The rest of the paper is organized as follows. Section 2 includes the methodology of the proposed work. Section 3 presents the performance evaluation of each regression model. Finally, the conclusion is provided in Sect. 4.

2 Methodology

Identifying the most important biomarkers and predicting the cognitive score help clinicians to identify AD progression at an early stage. This work utilizes the structural MR image features in the ADNI dataset [9] to identify the biomarkers and predict the cognitive scores using the machine learning models. The proposed framework is given in Fig. 1, which consists of (1) preprocessing, (2) modelling, (3) optimization and (4) cognitive score prediction.

2.1 Data Preprocessing

The quality of the data used for processing has a significant impact on the accuracy of cognitive score estimation. ADNI dataset is developed by the researchers by collecting, validating and utilizing data from different modalities and cognitive tests for a long period of time. As a result, it has various limitations, such as random missing information owing to irregular or failure in follow-up visits, which must be addressed effectively in order to enhance overall estimation accuracy. This leads to the necessity of preprocessing of data collected from ADNI dataset. One of the major drawbacks of ADNI dataset is the missing values [10] that can occur due to several reasons. This can be addressed by the elimination of corresponding entities or by the estimation of missing values from the available data. The entity elimination approach may not give promising results due to the lack of sufficient data to be processed after the elimination. Hence, the estimation of missing values becomes more relevant in practical applications. As the accuracy of estimating missing information increases, the overall accuracy of the prediction also increases. The most challenging aspect of this work is selecting the most appropriate technique from the numerous available methods to fill the missing values that are non-uniformly spread in the clinical data. Wrong estimation of the missing value can

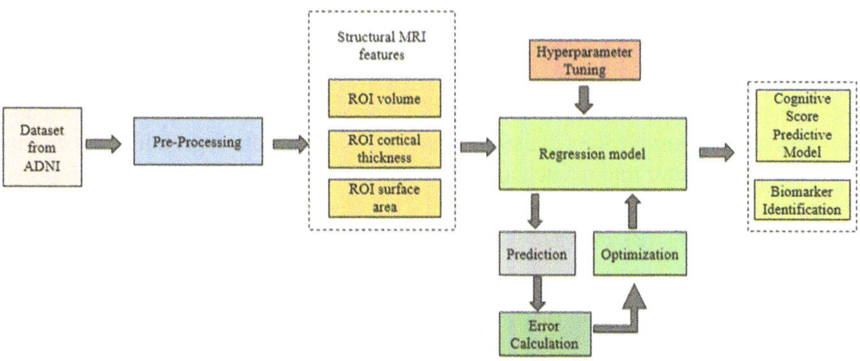

Fig. 1 Framework of the proposed design

lead to erroneous cognitive score predictions, which can lead to misinterpretation of the disease stage. A method that uses proximity imputation to identify the pattern of the missing value can be a suitable choice for clinical application. Hence, random forest (RF) algorithm is used to fill the missing observations [11]. The number of missing values associated with each feature varies, and the sequence in which available missing fields are treated has an impact on the missing value estimate. The estimation error can be minimized if the missing values are identified in a progressive manner from the features with minimum number of missing information to maximum number of missing information. Hence, as the first step, the features are ranked in ascending order based on the number of missing values. For imputation, the least number of missing field features is initially chosen and filled with rough values. The proximity matrix is then constructed using the RF-based imputation process to recalculate the missing value, which improves the outcome. It is observed that values associated with the different features in the dataset belong to different scales. Hence, min-max normalization is applied on the data which rescales all the features to the range of [0,1].

From the preprocessed data, the structural properties of the brain, such as volumes, cortical thickness and surface areas of ROIs [12], which have the potential to predict the subject's cognitive score, are considered for further modelling and analysis. To reduce the overall complexity and to improve the prediction accuracy, a study is conducted to identify the most important ROIs that influence accurately the prediction of the cognitive score. The next section describes the construction of predictive models from the selected features and its training and the testing process.

2.2 Modelling

The predictive modelling aims to predict the score based on the features of the structural MR images obtained from the ADNI. The predictive model is a machine learning based mathematical model that learns the pattern from the input features during the training phase. Other than the learnable parameters of the model, the hyperparameter, which is not directly learnable from the features, is initialized before the training process. Hyperparameters are significant because they directly influence the training algorithm's behaviour and have a major impact on the model's performance. Out of the many regression algorithms available to train the model and to predict the cognitive score, the proposed study considers regression algorithms with hyperparameter tuning such as ridge [6], lasso [7] and elastic net [8] to improve the prediction accuracy. The cost functions of ridge, lasso and elastic net are given, respectively, in (1), (2) and (3).

$$\beta \min L_{\text{ridge}} = \frac{1}{2}\|Y - X\beta\|_2^2 + \lambda\|\beta\|_2^2 \qquad (1)$$

$$\beta \min L_{\text{lasso}} = \frac{1}{2}\|Y - X\beta\|_2^2 + \lambda\|\beta\|_1 \quad (2)$$

$$\beta^{\min} J(\beta) = \frac{\|Y - X\beta\|^2}{2n} + \lambda\left[\frac{(1-\alpha)}{2}\sum_{j=1}^{m}\beta_j^{\,2} + \alpha\sum_{j=1}^{m}\|\beta\|\right] \quad (3)$$

where X is the feature matrix of subjects, Y is the corresponding target vector, β is the coefficient vector, α is the mixing parameter, m is the number of features, n is the total number of data samples and λ is the regularization parameter. The overall performance of the model can be improved only by minimizing the cost function which can be possible through the optimization process.

2.3 *Optimization*

Coordinate descent [13] is used as the optimization algorithm which minimizes along each coordinate direction in a successive way to find the minimum value of the cost function. During each iteration, the algorithm determines a coordinate and then tries to minimize it while keeping all other coordinates fixed.

2.4 *Predicting Cognitive Scores*

The linear predictive function used to estimate the cognitive score of a subject using the regression models is given in (4).

$$Y_{\text{pred}} = X * \beta + \text{intercept} \quad (4)$$

where *X* represents the MRI features selected for a particular time point, Ypred indicates the corresponding predicted cognitive scores (MMSE or ADAS13) at that time point and β represents the coefficient matrix or the weight matrix of the selected features and *intercept* denotes the bias.

3 Result and Discussions

Experiments are conducted to predict the cognitive score by implementing the proposed model and training/testing it with the ADNI dataset. The ADNI dataset contains the longitudinal data of 356 subjects with 124 CN, 18 EMCI, 146 LMCI

Table 1 Demographic chart

Category	CN	EMCI	LMCI	AD
Number	124	18	146	68
Gender(M/F)	65 / 59	9 / 9	90 / 56	38 / 30
Age (y, avg ± std)	75.39 ± 5.28	71.43 ± 8.25	74.23 ± 7.11	74.73 ± 7.75
Education (y, avg ± std)	15.97 ± 3.02	15.22 ± 2.32	15.99 ± 3.16	15.03 ± 3.01

and 68 AD. The mean and standard deviation of the demographic information of all subjects are given in Table 1.

The baseline, 6th, 12th and 24th month longitudinal data of each subject are used in this study for the cognitive score prediction which are separately taken and split into training and testing data with percentage split of 0.3. Each input set consists of 356 subjects and their corresponding 318 MRI features obtained from each visit. The ridge, lasso and elastic net models are trained and evaluated.

For the quantitative performance evaluation of each model, mean absolute error (MAE), root mean squared error (rMSE) and Pearson correlation coefficient (CC) are employed as metrics and are represented in (5), (6) and (7), respectively.

$$\text{MAE}\left(y,\hat{y}\right)=\frac{1}{2}\sum\nolimits_{i=1}^{N}\left|y_i-\hat{y}_i\right| \tag{5}$$

$$r\text{MSE}\left(y,\hat{y}\right)=\sqrt{\frac{1}{N}\sum\nolimits_{i=1}^{N}\left(y_i-\hat{y}_i\right)^2} \tag{6}$$

$$CC\left(y,\hat{y}\right)=\frac{\text{cov}\left(y,\hat{y}\right)}{\sigma(y)\sigma\left(\hat{y}\right)} \tag{7}$$

where y is the ground truth of target, $\hat{y}$ is the value predicted by the model, N is the number of data samples, cov is the covariance and σ is the standard deviation.

The performance evaluation of ridge, lasso and elastic net models based on MAE in MMSE and ADAS13 predictions is shown in Tables 2 and 3. rMSE in MMSE and ADAS13 predictions obtained from ridge, lasso and elastic net models are given in Tables 4 and 5. From the analysis it is observed that the prediction error is less in elastic net model at baseline, 6th and 12th compared with Ridge and Lasso models. Tables 6 and 7 show the performance evaluation of ridge, lasso and elastic net models based on CC in MMSE and ADAS13 prediction. Figures 2, 3 and 4 show the scatter plots of actual vs predicted MMSE scores of regression models for baseline, 6th and 12th month data, respectively. Figures 5, 6 and 7 show the scatter plot of actual vs predicted ADAS13 scores of regression models for baseline, 6th and 12th month data, respectively. The cognitive scores of baseline, 6th and 12th month

Table 2 MAE performance of models in MMSE prediction

	Baseline	6th	12th
Ridge	0.192398503	0.137685446	0.128832169
Lasso	0.165793355	0.116397636	0.108441351
Elastic net	0.163581701	0.11555035	0.104943214

Table 3 MAE performance of models in ADAS13 prediction

	Baseline	6th	12th
Ridge	0.129369518	0.140464001	0.130050318
Lasso	0.114829927	0.123100919	0.120424193
Elastic net	0.113464207	0.120831327	0.117705317

Table 4 rMSE performance of models in MMSE prediction

	Baseline	6th	12th
Ridge	0.241091403	0.177523904	0.166202076
Lasso	0.213991455	0.152872087	0.150484603
Elastic net	0.211315979	0.151530291	0.145963517

Table 5 rMSE performance of models in ADAS13 prediction

	Baseline	**6th**	**12th**
Ridge	0.170180695	0.18042717	0.167662843
Lasso	0.151315473	0.161297601	0.160821359
Elastic net	0.15020396	0.159886084	0.160071134

Table 6 CC performance of models in MMSE prediction

	Baseline	6th	12th
Ridge	0.40024053	0.4703327	0.55950827
Lasso	0.4907158	0.57879778	0.66238216
Elastic net	0.50958975	0.58635744	0.68713715

Table 7 CC performance of models in ADAS13 prediction

	Baseline	6th	12th
Ridge	0.50379534	0.49319093	0.62327193
Lasso	0.60434986	0.5819976	0.6627355
Elastic net	0.61401137	0.59378268	0.66749519

data are predicted using the elastic net model. These predicted values can be used as features for the cognitive score prediction of 24th month data using the elastic net model. Table 8 shows the performance evaluation of 24th month prediction of MMSE and ADAS13 scores respectively. Figure 8a and 8b show the scatter plots of the actual and estimated value for 24th month data for MMSE and ADAS13 scores, respectively.

Table 9 lists the top 10 ROIs selected by the elastic net model for cognitive prediction. These brain regions are affected by Alzheimer's and aid in the prediction of MMSE and ADAS13 scores for longitudinal data. Table 10 shows the strong correlation of the top 3 brain regions selected by the elastic net model with the MMSE and ADAS13 scores, respectively, revealing that the elastic net model has successfully identified the important biomarkers for AD detection.

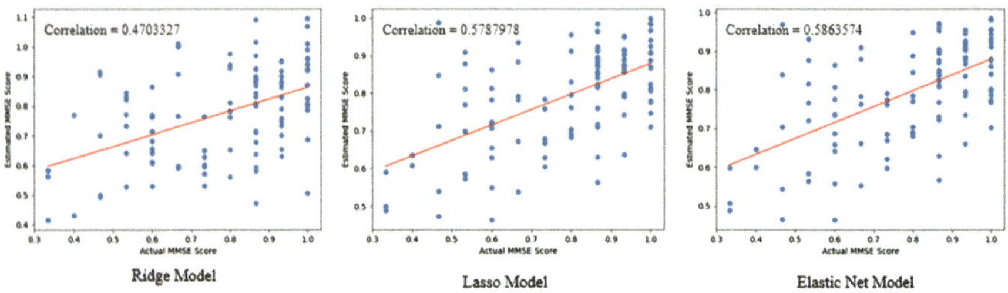

Fig. 2 Scatter plots of the actual MMSE score versus predicted values on testing data achieved by regression models for baseline data

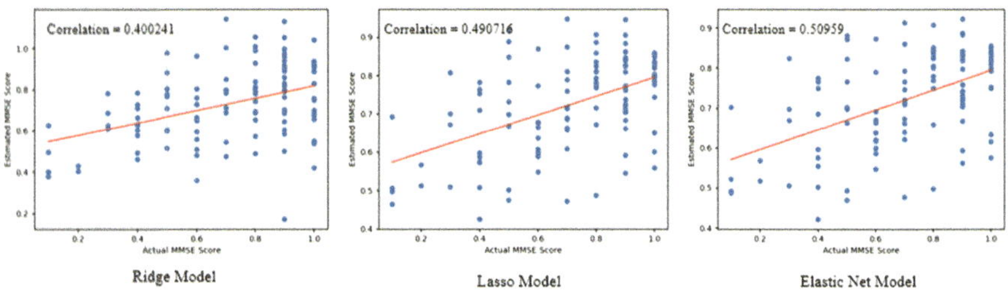

Fig. 3 Scatter plots of the actual MMSE score versus predicted values on testing data achieved by regression models for 6th month data

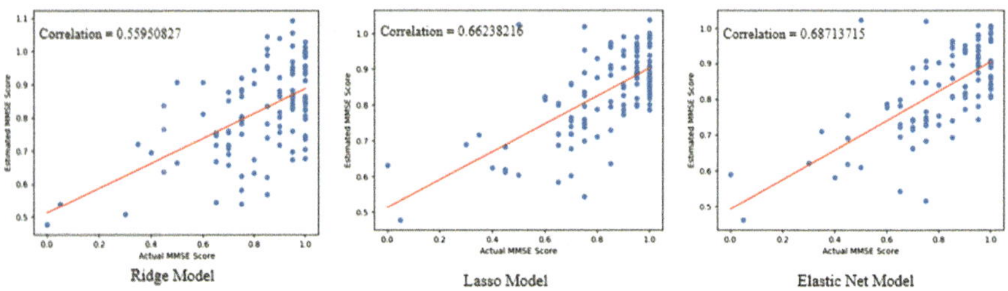

Fig. 4 Scatter plots of the actual MMSE score versus predicted values on testing data achieved by regression models for 12th month data

Figures 9, 10 and 11 show the correlation of these biomarkers with MMSE score. Figures 12, 13 and 14 show the correlation of these biomarkers with the ADAS13 score.

4 Conclusion

Detection of cognitive issues that aids in the diagnosis of AD is possible through cognitive tests which can be biased by several factors. The study is conducted using longitudinal MRI data from ADNI to predict cognitive scores at the corresponding

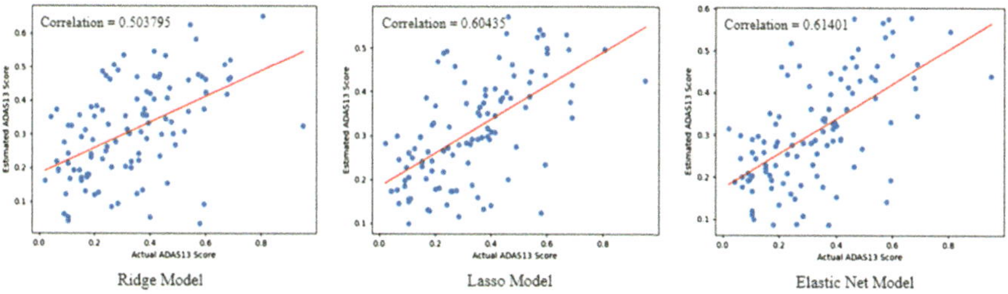

Fig. 5 Scatter plots of the actual ADAS13 score versus predicted values on testing data achieved by regression models for baseline data

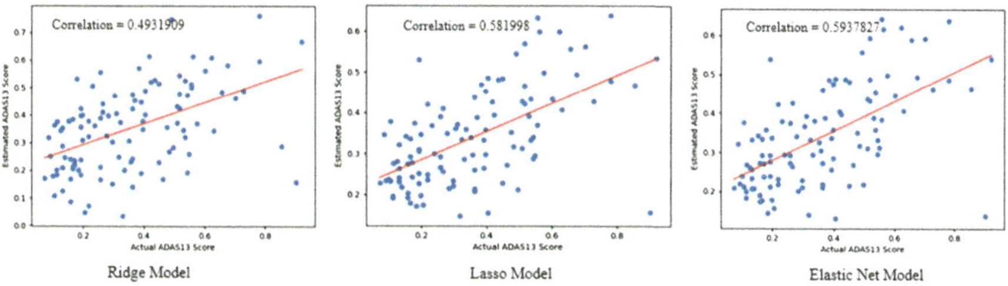

Fig. 6 Scatter plots of the actual ADAS13 score versus predicted values on testing data achieved by regression models for 6th month data

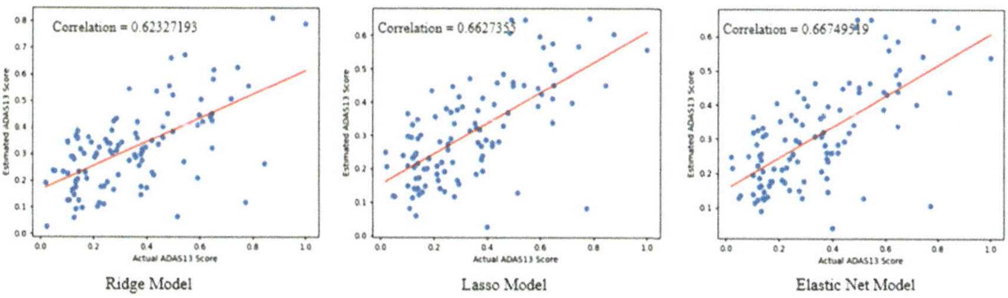

Fig. 7 Scatter plots of the actual ADAS13 score versus predicted values on testing data achieved by regression models for 12th month data

Table 8 Performance in MMSE and ADAS13 prediction of 24th month data

	MAE	RMSE	CC
MMSE	0.09942	0.13763252	0.653
ADAS13	0.11752	0.1562791	0.668

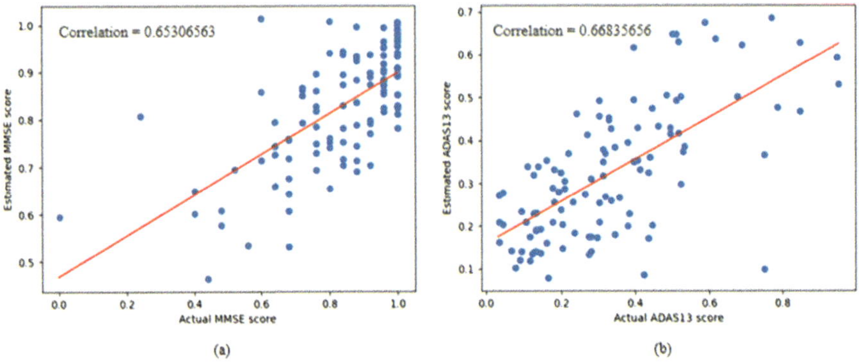

Fig. 8 (**a**) Scatter plot of the actual and estimated MMSE for 24th month data. (**b**) Scatter plot of actual and estimated ADAS13 for 24th month data

Table 9 Top 10 ROI

MMSE	ADAS13
Left hippocampus	Left hippocampus
Left inferior temporal	Left inferior temporal
Right entorhinal	Right entorhinal
Left entorhinal	Left entorhinal
Left middle temporal	Left inferior parietal
Left parahippocampal	Left parahippocampal
Left fusiform	Left fusiform
Right paracentral	Left precuneus
Left inferior parietal	Left cerebellum cortex
Left inferior lateral ventricle	Left inferior lateral ventricle

Table 10 CC of ROIs with MMSE and ADAS13 scores

Score	ROI	Baseline	6th	12th
MMSE	Left hippocampus	0.4916945	0.517639	0.51944868
	Left inferior temporal	0.421935	0.519561	0.56115
	Right entorhinal	0.4786033	0.544555	0.53288279
ADAS13	Left hippocampus	−0.513364	−0.557095	−0.583707
	Left inferior temporal	−0.488303	−0.551591	−0.589483
	Right entorhinal	−0.509681	−0.542425	−0.606489

time point. The proposed study aimed at the accurate estimation of cognitive scores such as MMSE and ADAS13 from the structural features of brain MRI using regression models such as the ridge, lasso and elastic net. The experiment is conducted using ridge, lasso and elastic model, a model that incorporates the qualities of the ridge and lasso models by combining L1 and L2 regularization. It is observed that the features selected by elastic net model are found to be highly correlated with the cognitive scores. Hence, elastic net model has the lowest MAE, lowest rMSE and highest CC compared to ridge and lasso models. MMSE and

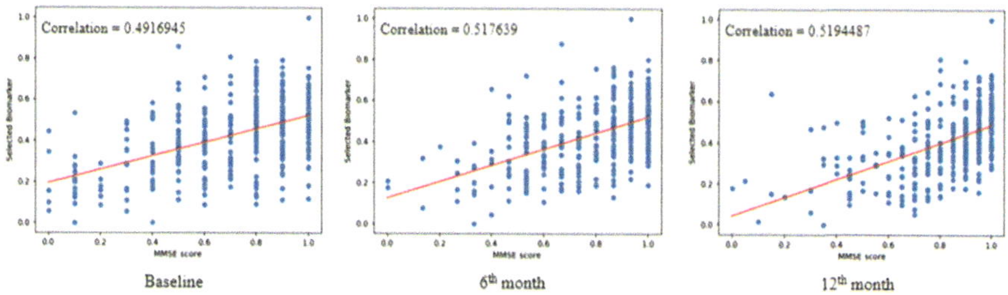

Fig. 9 Correlation between left hippocampus biomarker and MMSE score

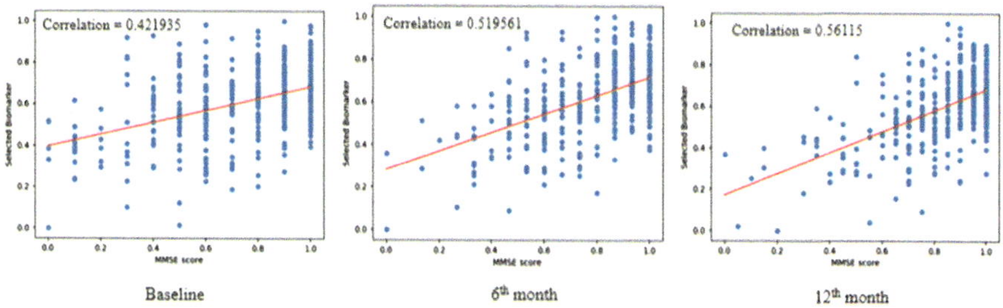

Fig. 10 Correlation between left inferior temporal biomarker and MMSE score

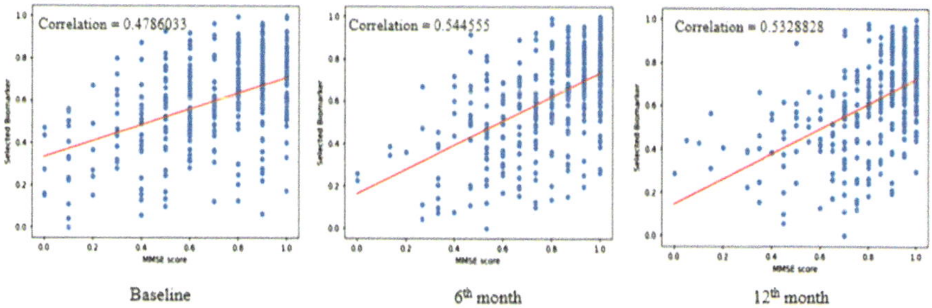

Fig. 11 Correlation between right entorhinal biomarker and MMSE score

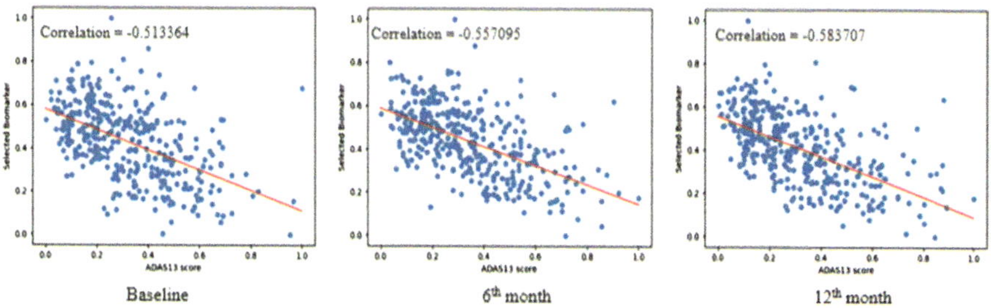

Fig. 12 Correlation between left hippocampus biomarker and ADAS13 score

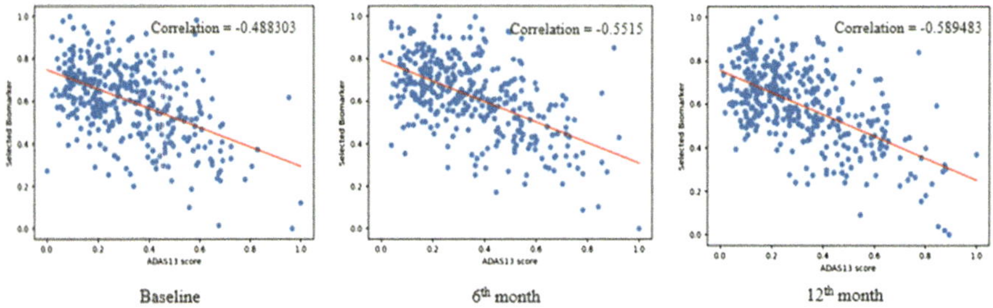

Fig. 13 Correlation between left inferior temporal biomarker and ADAS13 score

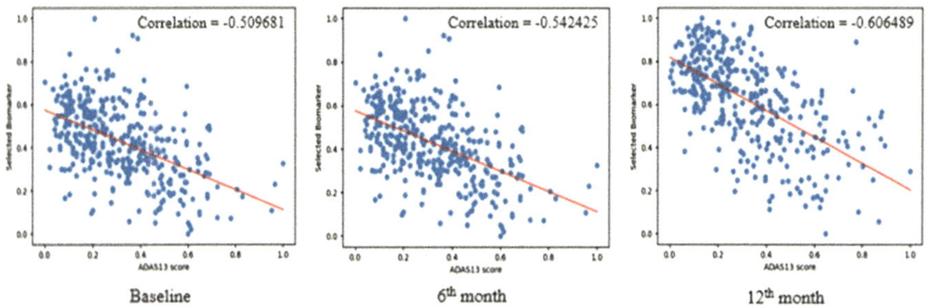

Fig. 14 Correlation between right entorhinal biomarker and ADAS13 score

ADAS13 scores predicted for baseline, 6th and 12th month data with elastic net model outperform other two models in terms of mean absolute error. MMSE and ADAS13 scores estimated by lasso for the 24th month using the predicted cognitive scores of baseline, 6th and 12th month provide accurate results. The elastic net model can be utilized as a practical model in cognitive score prediction and to identify important biomarkers for AD detection.

References

1. J. Rasmussen, H. Langerman, Alzheimer's disease–why we need early diagnosis. Degener. Neurol. Neuromuscul. Disease **9**, 123 (2019)
2. E. Grober, C.B. Hall, R.B. Lipton, A.B. Zonderman, S.M. Resnick, C. Kawas, Memory impairment, executive dysfunction, and intellectual decline in preclinical Alzheimer's disease. J. Int. Neuropsychol. Soc. **14**(2), 266–278 (2008)
3. D. Berron, D. van Westen, R. Ossenkoppele, O. Strandberg, O. Hansson, Medial temporal lobe connectivity and its associations with cognition in early Alzheimer's disease. Brain **143**(4), 1233–1248 (2020)
4. E. Devenney, J.R. Hodges, The mini-mental state examination: Pitfalls and limitations. Pract. Neurol. **17**(1), 79–80 (2017)
5. J. McDowd, L. Hoffman, E. Rozek, K.E. Lyons, R. Pahwa, J. Burns, S. Kemper, Understanding verbal fluency in healthy aging, Alzheimer's disease, and Parkinson's disease. Neuropsychology **25**(2), 210 (2011)

6. C. Westbrook, J. Talbot, *MRI in Practice* (Wiley, 2018)
7. I. Arevalo-Rodriguez, N. Smailagic, Marta Roqué i Figuls, A. Ciapponi, E. Sanchez-Perez, A. Giannakou, O.L. Pedraza, X.B. Cosp, S. Cullum, Mini-Mental State Examination (MMSE) for the detection of Alzheimer's disease and other dementias in people with mild cognitive impairment (MCI). Cochrane Database Syst. Rev. (3) (2015)
8. S. Iddi, D. Li, P.S. Aisen, M.S. Rafii, W.K. Thompson, M.C. Donohue, Predicting the course of Alzheimer's progression. Brain Inform. **6**(1), 1–18 (2019)
9. ADNI homepage http://adni.loni.usc.edu/
10. S. Campos, L. Pizarro, C. Valle, K.R. Gray, D. Rueckert, Héctor Allende.: Evaluating imputation techniques for missing data in adni: A patient classification study, in *Iberoamerican Congress on Pattern Recognition*, (Springer, Cham, 2015), pp. 3–10
11. A. Thushara, C.U. Amma, A. John, R. Saju, Multimodal MRI based classification and prediction of Alzheimer's disease using random forest ensemble, in *2020 Advanced Computing and Communication Technologies for High Performance Applications (ACCTHPA)*, (IEEE, 2020), pp. 249–256
12. R.V. Marinescu, N.P. Oxtoby, A.L. Young, E.E. Bron, A.W. Toga, M.W. Weiner, F. Barkhof, N.C. Fox, S. Klein, D.C. Alexander, Tadpole challenge: Prediction of longitudinal evolution in Alzheimer's disease. arXiv preprint arXiv: 1805.03909 (2018)
13. S.J. Wright, Coordinate descent algorithms. Math. Program. **151**(1), 3–34 (2015)

An Automated Cervical Cancer Detection Mechanism Using Pap Smear Images

K. Geetha, S. Aarthi, N. Sasikaladevi, and C. Mala

Abstract Cervical cancer is one predominant type of cancer and ranks as the second most common type of cancer that often affects women over 30 years. Among all cancers in women, 6–29% constitutes cervical cancer. The test used for detecting the tumor is called the Pap smear test, which requires expert pathologists to classify the abnormal cells in the precancerous earlier stages. In rural villages, such experts are not easily accessible, and they need emerging technologies for the early detection of precancerous lesions. This proposed model is constructed using a fully connected 24 × 1 CNN model in the Herlev dataset of Pap smear cell images. This automated pap smear image classification model works through the learning and analysis of fully connected CNN with data augmentation process. Experimental results of the proposed scheme claim that this model is classifying the abnormal cervical cancer cells with an accuracy of 90.2%.

Keywords Cervical cancer · Pap smear cell · Medical image analysis · Medical image classification · Data augmentation · Fully connected 24 × 1 CNN and Herlev dataset

1 Introduction

Cancer is caused by the irresistible growth of abnormal cells in any biological structure that can spread into the entire human body. Worldwide, cervical cancer ranks fourth among all the common cancer types in women based on the statistics

K. Geetha · S. Aarthi (✉) · N. Sasikaladevi
CSED, SoC, SASTRA Deemed University, Thanjavur, Tamil Nadu, India
e-mail: geethavalavan@gmail.com; aarthiresearch047@gmail.com; sasikalade@gmail.com

C. Mala
Department of Computer Science and Engineering, National Institute of Technology, Tiruchirappalli, Tamil Nadu, India
e-mail: mala@nitt.edu

A. Haldorai et al. (eds.), *4th EAI International Conference on Big Data Innovation for Sustainable Cognitive Computing*, EAI/Springer Innovations in Communication and Computing, https://doi.org/10.1007/978-3-031-07654-1_18

provided by the World Health Organization. In 2018, over 5.7 lakh women were diagnosed with cervical cancer. Over 3 lakh women lost their lives due to this disease. But this disease is cured by effective screening, early detection, and diagnosis at the precancerous stages. According to the statistical report, more than 60% of women are unaware of cervical cancer, and the remaining 40% are aware of the risk. So there is a requisite to identify and treat the precancerous lesions to reduce the fatality rate caused by cervical cancer in women. For detecting these precancerous lesions, the HPV test and Pap test need to examine the cervical lesions caused by human papillomaviruses (HPV). The cervical cells are examined through the HPV test for detecting the severity of infection. Then the Pap test [1] collects the cervical cells and examines any changes present in each cell division. This Pap test is highly needed to detect early precancerous lesions. The results will be normal cells and abnormal cells.

Considering the significance of deep learning algorithms, several different models have been utilized to facilitate various researches in biomedical technologies. These algorithms eliminate medical errors by training with larger datasets and better accuracy. Such algorithms are defined to understand the medical images to identify whether the images represent any health-related abnormalities or not. Analyzing various algorithms, convolutional neural network (CNN) is the prime focus for processing (training and testing) large datasets [2] in the health sector for predicting the presence of diseases such as cancer, tumor, diabetes, and various health-related issues. In the deep learning algorithm, medical images are given as input to the model and output as a binary classification to classify the presence or absence of the diseases. But, detecting cervical cancer predominantly in the form of precancerous lesions is very difficult. So, deep learning models like CNN are significant to provide better results in terms of accuracy.

The purpose of the proposed work is to promote an automated tool for classifying Pap smear cells (single cell) with higher accuracy. The *Automation* is required for the cervical cancer detection process due to the high severity rate and high mortality rate in the women worldwide. This automation can be achieved through the fully connected CNN for the depth feature analysis and selection process in Pap smear images. This automated interface is implemented for satisfying various properties including easy access, reliability, and better response time. A good interface is essential for providing great accessibility and ease of use in operation for any novice users in places where expert pathologists are rare. The second crucial objective of the work is to build the convolution neural networks (CNN) layers and fine-tune the parameters [3] for better results.

1.1 Motivation and Contribution

There are numerous computer-assisted methods are available for the medical image analysis and classification. But, this cervical cancer lesion screening and analysis is more tedious when compared with any other medical image classification. The detection of lesions in Pap smear cells is not cost-effective. There must be a need for the processes like image acquisition, preprocessing, segmentation, and feature extraction process in Pap smear image analysis. Now a day, the deep learning technologies provide a better platform for medical image analysis and segmentation. After analyzing the severity level of this cervical cancer in women, the motivation of this paper is to propose an interface formulated on a deep learning neural network that can classify a given Pap smear cell image as an abnormal cell or a normal cell. For making the CNN effective, Data augmentation process is utilized to learn the features from all the image dimensionalities. This model purportedly implements the semantic segmentation using the 24 $\times$ 1 fully connected CNN layer. And this model is cost-effective when compared with the models with fuzzy logic and machine learning algorithms. This proposed model can help physicians to identify abnormal cervical cells, and this procedure can be done and examined even through remote health care monitoring systems (RHMR).

1.2 Paper Organization

The rest of the remaining paper is organized as follows. Section 2 provides challenges faced by the related works done in this Pap smear cell classification. Section 3 illustrates a summary of the proposed CNN scheme with the architecture and workflow. In Sect. 4, we analyze the implementation and results of the proposed model with various performance metrics. At last, we conclude this study with the future research opportunities in Sect. 5.

2 Related Works

According to Dong et al. [4], binary classification of the cervical cancer cell is implemented using an SVM classifier in which cell feature extraction is processed through classification and regression trees (CART). Feature extraction depends on color, textures, and morphological features. In this paper, the CART algorithm provides the feature selection, and then PSOSVM (particle swarm optimization-support vector machine) is used as the classifier. But the computational cost for extracting the features from Pap smear single-cell images will be high. For improving the classification of precancerous lesions, cancer lesion segmentation and Bethesda system (ensemble technique) is used in [5]. The higher loss of the model will impact the efficacy of the Bethesda system. For eliminating this limitation, Fourier-transform infrared (FTIR) data [6] with support vector machines (SVM) classifier is used for classification that results in an accuracy of 72%. But the accuracy is comparatively low when compared with the existing schemes. Voting and DNN classifiers [7] are devised based on a voting classifier and deep neural networks (DNN). If the cytology and biopsy datasets are used for evaluating the performance metrics, this model provides an automated feature extraction using large datasets. But it does not give better accuracy for various types of cancerous cells.

L. Zhang [8] focused on the ConvNet model, which is used for the classification of pap smear cells on Herlev dataset and HEMLBC liquid-based cytology datasets. There is a requirement for a better segmentation model due to the poor accuracy. A semi-automatic analysis of PC-based cellular images [9] is used for the segmentation of nuclear, and cytoplasm in the cell images is suggested, and textural features are added in this to implement the SVM classifier. In [10], image classification is carried out using Efficient Human Epithelial-2 (HEp-2) by implementing convolutional neural networks (CNN), and this implementation of the deep learning method is expected to produce better results. In paper [11], various shape-based features were calculated for the nucleus and the cytoplasm. This model is concerned only with shape-based features like area, perimeter, and eccentricity. After calculating the above features for the nucleus and cytoplasm, the N/C ratio is calculated as a measurement to distinguish the cell as normal and abnormal. For a normal cell, the ratio is small, and the nucleus is circular. For an abnormal cell, the N/C ratio surges the shape of the nucleus changes to the elliptical. Feature selection is done using SVM-RFE that removes the insignificant features. The insight of building fully convolutional networks that can be implemented in different applications is explained in [12].

A report on a powerful active contour segmentation model called the Chan-Vese model [13] explains the complete implementation methodology using level set formulation. Here, [14, 15] contour is chosen, and a signed distance function is defined. After verifying the magnitude of this function, the sign is decided based on the location (outside-negative, inside-positive, on the contour-zero).

Based on analyzing the related works done, the impact of the morphological feature in the binary classification of Pap smear cell images is highly induced in the proposed research. Fully convolutional networks depend on the input in the local spatial dimension. This CNN is technically used for the semantic segmentation of the images to avoid the difficulties in the image segmentation process that is even highly tedious in the case of medical images.

3 Proposed Methodology

The Pap smear images are collected through the Pap test, and a single-cell image is extracted from multiple cell images. Then, the accuracy is improved by the augmentation of the Pap smear cell images present in the dataset. These augmented images are given as an input to the fully connected convolutional neural network (CNN) for identifying the abnormality in Pap smear cell images.

It (see Fig. 1) illustrates the general workflow for classifying the Pap smear cell images to the appropriate binary class such as normal or abnormal. First, Pap smear cell images are collected from the Pap test, and the single-cell image of a pap

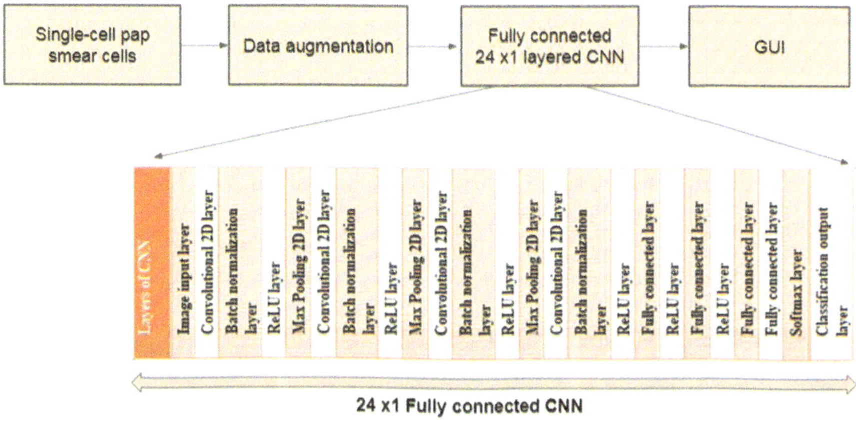

Fig. 1 General workflow for the identification of classes in Pap smear cell images

smear is extracted from the multiple cells. Next, the corresponding unit cell image is preprocessed for better classification. This model includes the segmentation process and is finally passed into the 24 × 1 convolutional neural networks. Then this model segments the Pap smear cell images semantically and classifies them into the appropriate classes. Here, a component of images that are fed into the CNN is known as channels. These channels represent the image in red, green, and blue with pixel values from 0 to 255 and stacked one over another. If the classification of Pap smear cells is resolved in a machine learning algorithm, it needs separate feature extraction and classifier for identifying the abnormality in Pap smear cell images. But this problem is resolved by providing CNN for classification.

3.1 Dataset

DTU/Herlev Pap smear benchmark dataset (7 class dataset) is used in this work. This dataset consists of 917 single-cell images of Pap smear cells. There are seven classes present in this dataset. *The normal cells are defined in the first three classes of cells, and the abnormal cells are listed in the last four classes of cells in* Table 1. Each class is described in Table 1.

Figure 2 illustrates the various sample images of Pap smear cell images. Each consists of different shapes, sizes, and morphological features. For instance, the size of the nucleus and nucleoplasm is not the same for all the cells. They might differ based on color, shape, and deformity. There are various types of cells introduced in this dataset tabulated in Table 1. The size of the nucleus and its shape and texture will provide an impact on the results. And other factors like nucleoplasm are evaluated to improve the classification model.

Table 1 Herlev dataset

Serial number	Types of cells	Size of the nucleus	Count
1	Superficial squamous epithelial	**Very small**	**97**
2	Intermediate squamous epithelial	**Large**	**70**
3	Columnar epithelial	**Large/light**	**74**
4	Mild dysplasia	**Medium**	**182**
5	Moderate dysplasia	**Large/dark**	**146**
6	Severe dysplasia	**Large/dark/deform**	**197**
7	Carcinoma in situ	**Large/dark/deform**	**150**

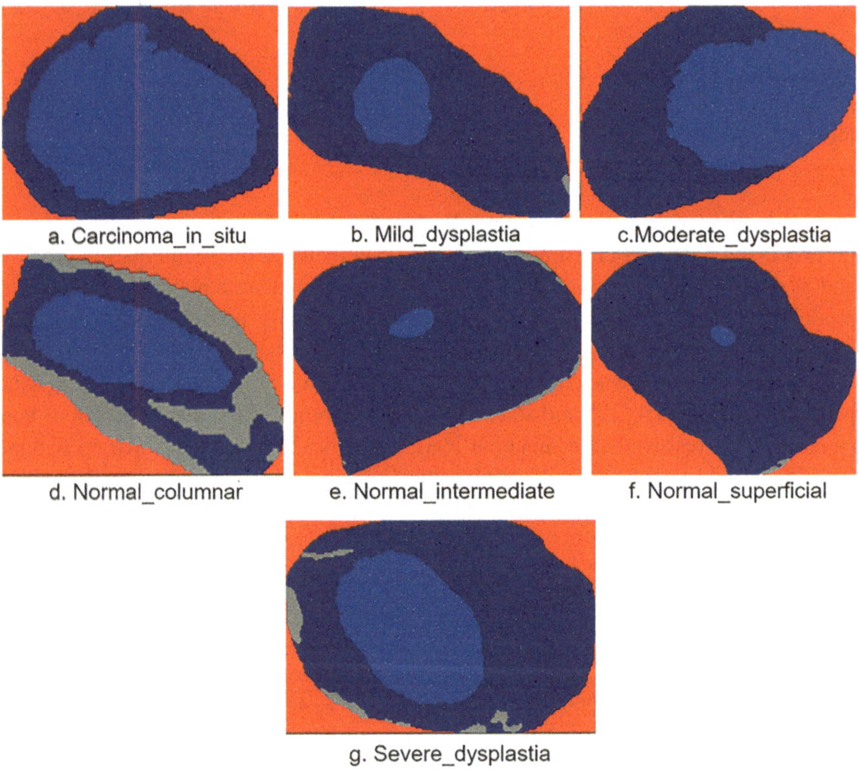

Fig. 2 Sample Pap Smear cell images

3.2 Implementation

The convolutional neural network (CNN) model is implemented using the MATLAB platform by deep network analyzer tool. Figure 3a represents the healthier and unhealthier Pap smear cells, whereas Fig. 3b represents the abnormality in the morphological features of pap smear cells (see Fig. 3).

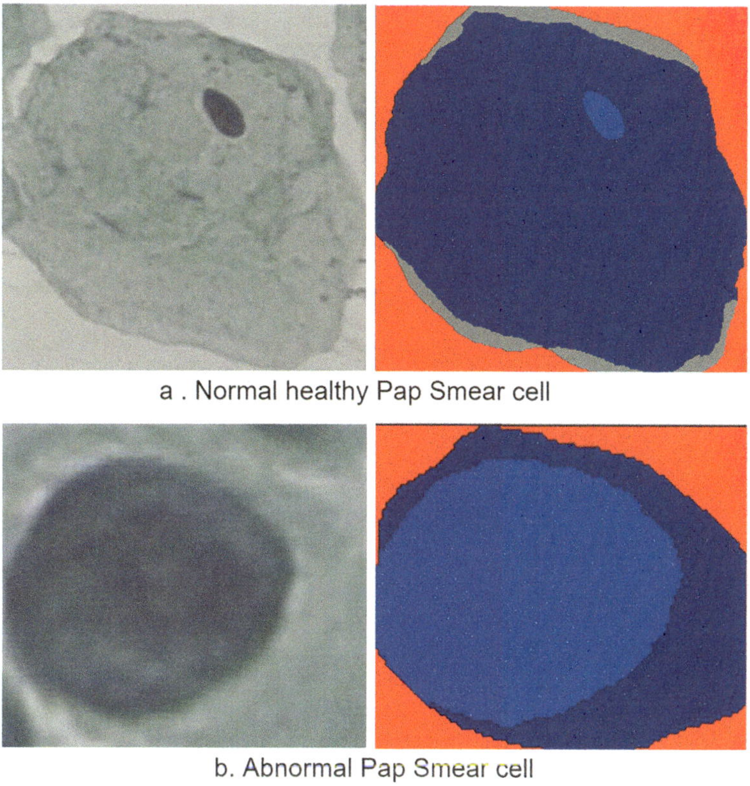

Fig. 3 (**a** and **b**) Representation of Healthy and Unhealthy Pap smear cells

First, data augmentation is used to improve the training model. This process includes image rotation, scaling, and reflection of the biomedical images. Next, these augmented images are passed to the convolutional layer. This convolutional layer learns the features of the input images. Then the resultant images are passed into the pooling layer that reduces the dimensionality of the feature maps and denoising those images. And then Max Pooling layer is required for selecting the maximum value of the portion of images. Finally, it is passed to the ReLu layer to which substitutes 0 for the negative pixel value. It is an element wise operation. The layers of CNN are analyzed using the deep network analyzer tool present in MATLAB. These iterated 24 layers are responsible for the above operations like convolution, ReLu, Batch normalization, and Max Pooling. As the result, the layers of CNN do not face any difficulties or errors in the implementation. The complete list of CNN is tabulated in Table 2.

Table 2 Lists of layers used in the proposed scheme

Serial number	Layers of CNN	Size
1	Image input layer	1×1
2	Convolutional 2D layer	1×1
3	Batch normalization layer	1×1
4	ReLU layer	1×1
5	Max Pooling 2D layer	1×1
6	Convolutional 2D layer	1×1
7	Batch normalization layer	1×1
8	ReLU layer	1×1
9	Max Pooling 2D layer	1×1
10	Convolutional 2D layer	1×1
11	Batch normalization layer	1×1
12	ReLU layer	1×1
13	Max Pooling 2D layer	1×1
14	Convolutional 2D layer	1×1
15	Batch normalization layer	1×1
16	ReLU layer	1×1
17	Fully connected layer	1×1
18	ReLU layer	1×1
19	Fully connected layer	1×1
20	ReLU layer	1×1
21	Fully connected layer	1×1
22	Fully connected layer	1×1
23	Softmax layer	1×1
24	Classification output layer	1×1

3.3 Algorithm

Step 1 First, Pap smear cell images are collected and stored in the large dataset, and the unit single cell is extracted from the multiple cell images. For this, we have to use the Herlev dataset that consists of different single-cell images extracted from the multiple cells with various morphological features in single-cell images.

Step 2 Data augmentation of the image dataset takes place through various operations like scaling the dimensions of medical images, rotating the input images, and reflecting the input images using the functions such as RandXScale, RandYScale, RandRotation, RandXReflection, and RandYReflection.

Step 3 The augmented images are passed into the layers of CNN which is 24 x1 proposed models in which the epoch hyper parameter value is set to 100. The initial learning rate value is considered to be a tuning parameter that is used to determine the step size. The entire layers present in CNN are listed in Table 2. The functions like convolution, batch normalization, ReLU, Max Pooling, and softmax are performed in this section to train the model with better accuracy. Here the base learning rate is set to 0.0100. Verbose is set to true that shows the entire progress of training over many iterations.

```
options = trainingOptions('sgdm', 'MaxEpochs',100,
'InitialLearnRate',0.01, 'Verbose',true, 'Plots',
'training-progress');
```

Step 4 Finally, GUI is designed for visualizing the testing images with greater accuracy that may be helpful for the exact prediction for the new input pap smear cell images.

```
I=imread("smear2005_big\A\148497551-148497588-001.bmp");  //input
I2=imresize(I,[64,64],'nearest');     //resize theimage
[Pred,scores]=classify(net,I2);             //predict theoutput
```

These are the steps followed in this proposed scheme to identify the abnormal Pap smear cell images using the data augmentation process and 24 x 1 CNN model.

4 Results and Discussions

This methodology of predicting the presence of cervical cancer on the pap smear cells that was carried out on Matlab provided a better accuracy of 90.2%. Table 3 represents the accuracy of different epochs running in a different iteration. Mini batch accuracy is improved when the epoch value and iteration increase. Here the base learning rate is maintained as 0.0100. And the loss is reduced when we increase the value for iteration and epoch.

Figure 4 illustrates the performance graphs of the proposed model. This plot illustrates the training model using 24 × 1 CNN. The performance graphs include both accuracy and loss for the proposed scheme with 100 epochs and 500 iterations.

Table 3 Accuracy with different epoch values

Sl. no.	Epoch	Iteration	Time elapsed	Mini batch accuracy	Mini-batch loss	Base learning rate
1	1	1	00:00:04	69.53%	0.7101	0.0100
2	10	50	00:01:31	89.06%	0.2612	0.0100
3	20	100	00:02:58	89.06%	0.2291	0.0100
4	30	150	00:04:44	90.63%	0.2186	0.0100
5	40	200	00:06:15	93.75%	0.2060	0.0100
6	50	250	00:07:46	90.63%	0.1772	0.0100
7	60	300	00:09:20	91.41%	0.1972	0.0100
8	70	350	00:10:51	92.19%	0.1444	0.0100
9	80	400	00:12:22	90.63%	0.1925	0.0100
10	90	450	00:13:52	92.19%	0.1610	0.0100
11	100	500	00:15:24	92.97%	0.1519	0.0100

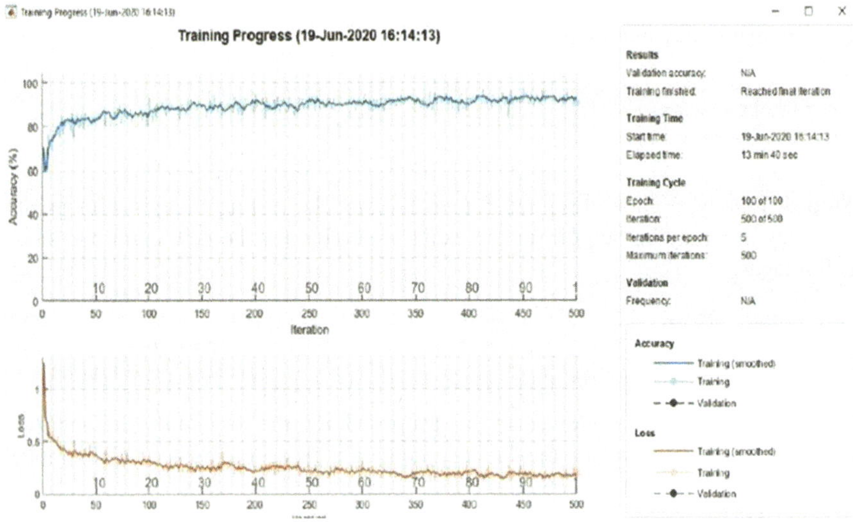

Fig. 4 Performance graphs for training phase

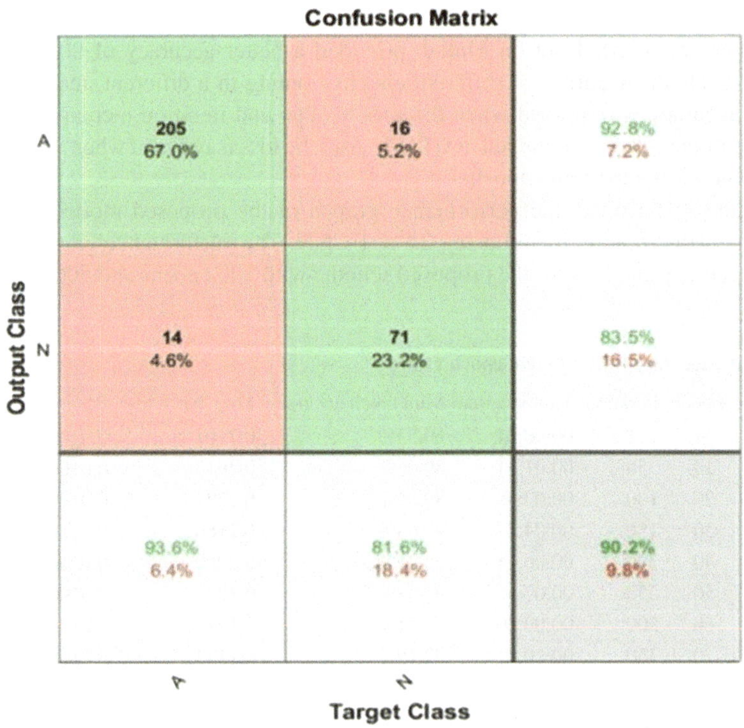

Fig. 5 Confusion matrix

Table 4 Performance metrics of the proposed scheme

Serial number	Measure	Value
1	Sensitivity	0.9358
2	Specificity	0.8169
3	Precision	0.9280
4	Negative predictive value	0.8345
5	False-positive rate	0.1831
6	False discovery rate	0.0720
7	False-negative rate	0.0642
8	Accuracy	0.9020
9	F1 score	0.9318

For this, the iteration per epoch is 5, and the total elapsed time is calculated as 13 min 40 sec.

Confusion matrix is defined for evaluating the performance of classification of cells into normal and abnormal illustrated in Fig. 5. The performance metrics of the proposed model is also tabulated (see Table 4). In Table 4, various metrics are calculated using the values of the confusion matrix.

Accuracy It is calculated as the percentage of correctly classified cell images into normal cells into the normal category (binary class- 0) and the abnormal cells into the abnormal cell category(binary class-1). In this model true positives are 71, false positives are 205, true negatives are 16, and false positives are 14.

$$\text{Accuracy} = \frac{TP + TN}{P + N}$$

Sensitivity It is the ratio of correctly identifying the abnormal cell images in this case. This model provides an accuracy of 90.2% with the total true negatives of 67% and true positives of 23.2%.

$$\text{Sensitivity} = \frac{TP}{TP + FN}$$

Specificity It is the ration of correctly identifying the normal cell images which must be listed in the false class.

$$\text{Specificity} = \frac{TN}{FP + TN}$$

Precision It represents the percentage of results obtained that are relevant. It helps to provide better results in imbalanced datasets. Herlev dataset is an imbalanced dataset that consists of 67% true negatives and only 23% as true positives.

$$\text{Precision} = \frac{TN}{FP + TP}$$

F1 Score It denotes the accuracy of the test. It takes into account both precision and recall.

$$F1 \text{ score} = \frac{2TP}{2TP + FP + FN}$$

Matthews Correlation Coefficient It represents the quality of classifications that are binary (having two classes). In this proposed scheme, the Matthews correlation coefficient value is 0.7576.

$$\text{Matthews correlation coefficient} = \frac{(TP * TN)\ (FP * FN)}{\sqrt{TP + FP} * (TP + FN) * (TN + FP)}$$

The GUI interface designed for the work is shown in Fig. 6. The following steps included in this interface are considered as below:

1. Select the cell image by clicking on the "browse" button. The entire path of the image file along with the image content will be displayed.
2. Select the "predict" button that displays the classification label of that image (abnormal or normal) along with accuracy.
3. Pap smear cell image that is passed as an input is classified into an abnormal cell with an accuracy of 99.404% and it is represented in the designated GUI Fig. 6.

Hence the proposed scheme will be better with higher accuracy rate when compared with the existing models illustrated in Table 5. The comparative analysis of the accuracy of existing model is given as below.

The classification approach using ANN and SVM for cervical cancer detection by extracting texture features has produced accuracy of 88.2%, sensitivity of 88.8%, and specificity of 85.9% on the Herlev dataset. This proposed model implementing on the same Herlev dataset has provided a significantly improved accuracy of about 90.2% and a sensitivity of 93.58%.

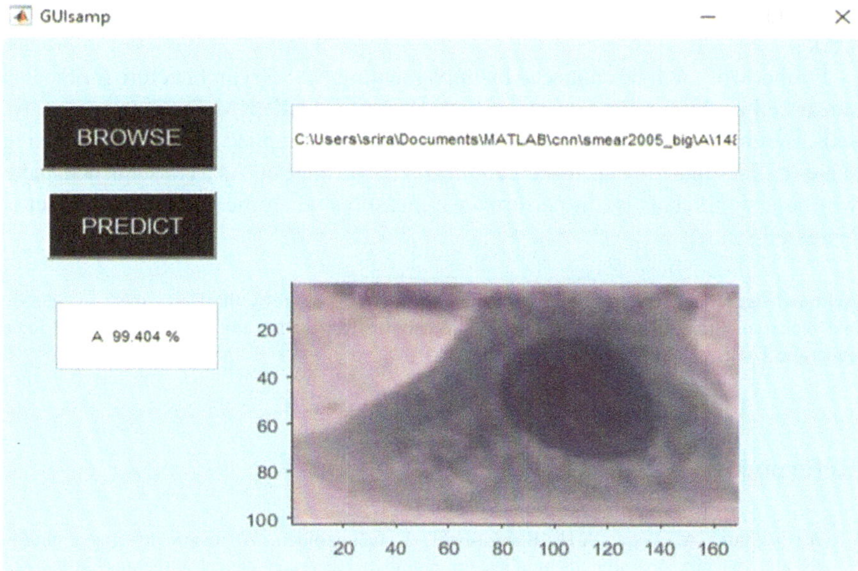

Fig. 6 GUI for classifying the Pap smear cell images

Table 5 Accuracy of various types of Pap smear cells in this GUI

Serial number	Types of cells	Class	Accuracy
1	Superficial squamous epithelial	**Normal**	**99.404**
2	Intermediate squamous epithelial	**Normal**	**90.456**
3	Columnar epithelial	**Normal**	**96.327**
4	Mild dysplasia	**Abnormal**	**91.563**
5	Moderate dysplasia	**Abnormal**	**98.343**
6	Severe dysplasia	**Abnormal**	**97.456**
7	Carcinoma in situ	**Abnormal**	**98.579**

5 Conclusion and Future Directions

Cervical cancer is considered to be the deadliest cancer if not identified in the early stages. This type of cancer requires a Pap test and HPV test for accurate prediction. However, expert medical examiners are essential for carrying out such tests efficiently that is difficult in highly populated rural areas of India. This automated system explained in this work can elegantly apply deep learning convolutional neural network (CNN) model to predict the abnormal and normal cells. This proposed work achieved 90.2% accuracy, 93.58% sensitivity, and 81.69% specificity. Results thus demonstrate that the above approach of implementing CNN on the Herlev dataset provides better results by examining the various performance metrics like accuracy, sensitivity, and specificity. Extracting the single-cell images from the multiple-cell image is a tedious process. So, considering the multiple cell

Pap smear images as the testing and training input will be imparted in our future work.

Further work will be enhanced by implementing NASNet architecture for the Pap smear cell analysis with datasets of both Herlev and SIPaKMeD (Cytology analysis). Moreover, many segmentation methods can be applied in the preprocessing stages to fine-tune the Pap smear cell images of the dataset. As a consequence, there will be a significant rise in performance measures when the size of the dataset is increased.

Acknowledgments This part of this research work is supported by the Department of Science and Technology (DST), the Science and Engineering Board (SERB), and the Government of India under the TARE grant (TAR/2020/000259).

References

1. W. William, A. Ware, A.H. Basaza-Ejiri, J. Obungoloch, A review of image analysis and machine learning techniques for automated cervical cancer screening from pap-smear images. Comput. Methods Prog. Biomed. **164**, 15–22 (2018). https://doi.org/10.1016/j.cmpb.2018.05.034. Epub 2018 Jun 26
2. A.L. Bogale, T. Teklehaymanot, J. Haidar Ali, G.M. Kassie, Knowledge, attitude and practice of cervical cancer screening among women infected with HIV in Africa: Systematic review and meta-analysis. PLoS One **16**(4), e0249960 (2021). https://doi.org/10.1371/journal.pone.0249960
3. K. Li, Z. Lu, W. Liu, J. Yin, Cytoplasm and nucleus segmentation in cervical smear images using radiating GVF Snake. Pattern Recogn. **45**(4), 1255–1264., ISSN 0031-3203 (2012). https://doi.org/10.1016/j.patcog.2011.09.018
4. N. Dong, M. Zhai, L. Zhao, C. Wu, Cervical cell classification based on the CART feature selection algorithm. J Ambient. Intell. Human Comput. **12**, 1–13 (2020)
5. K. Bora, M. Chowdhury, L.B. Mahanta, M.K. Kundu, A.K. Das, Automated classification of Pap smear images to detect cervical dysplasia. Comput. Methods Prog. Biomed. **138**, 31–47 (2017). https://doi.org/10.1016/j.cmpb.2016.10.001
6. E. Njoroge, S.R. Alty, M.R. Gani, M. Alkatib, Classification of cervical cancer cells using FTIR data, in *2006 International Conference of the IEEE Engineering in Medicine and Biology Society*, (New York, 2006), pp. 5338–5341. https://doi.org/10.1109/IEMBS.2006.260024
7. K. Rayavarapu, K.K.V. Krishna, Prediction of cervical cancer using voting and DNN classifiers, in *2018 International Conference on Current Trends towards Converging Technologies (ICCTCT)*, (Coimbatore, 2018), pp. 1–5. https://doi.org/10.1109/ICCTCT.2018.8551176
8. L. Zhang, L. Lu, I. Nogues, R.M. Summers, S. Liu, J. Yao, DeepPap: Deep convolutional networks for cervical cell classification. IEEE J. Biomed. Health Inform. **21**(6), 1633–1643 (2017). https://doi.org/10.1109/JBHI.2017.2705583
9. Y. Chen, P. Huang, K. Lin, H. Lin, L. Wang, C. Cheng, T. Chen, Y. Chan, J.Y. Chiang, Semi-automatic segmentation and classification of pap smear cells. IEEE J. Biomed. Health Inform. **18**(1), 94–108 (2014). https://doi.org/10.1109/JBHI.2013.2250984
10. Z. Gao, L. Wang, L. Zhou, J. Zhang, HEp-2 cell image classification with deep convolutional neural networks. IEEE J. Biomed. Health Inform. **21**(2), 416–428 (2017). https://doi.org/10.1109/JBHI.2016.2526603
11. A.H. Mbaga, P. Zhijun, Pap smear images classification for early detection of cervical cancer. Int. J. Comput. Appl. **118**(7), 10–16 (2015)

12. J. Long, E. Shelhamer, T. Darrell, Fully convolutional networks for semantic segmentation. Proc. IEEE Conf. Comput. Vis. Pattern Recognit., 3431–3440 (2015)
13. A Report on 'The Chan-Vese Algorithm'. Available at https://arxiv.org/ftp/arxiv/papers/1107/1107.2782.pdf
14. W. William, A. Ware, A.H. Basaza-Ejiri, J. Obungoloch, Cervical cancer classification from pap-smears using an enhanced fuzzy C-means algorithm. Inform. Med. Unlocked **14**, 23–33., ISSN 2352-9148 (2019). https://doi.org/10.1016/j.imu.2019.02.001
15. M. Anousouya Devi, S. Ravi, J. Vaishnavi, S. Punitha, Classification of cervical cancer using artificial neural networks. Procedia Comput. Sci. **89**, 465–472., ISSN1877-0509 (2016). https://doi.org/10.1016/j.procs.2016.06.105

Index

A. Haldorai et al. (eds.), *4th EAI International Conference on Big Data Innovation for Sustainable Cognitive Computing*, EAI/Springer Innovations in Communication and Computing, https://doi.org/10.1007/978-3-031-07654-1

The manufacturer's authorised representative in the EU is Springer
Nature Customer Service Centre GmbH, Europaplatz 3, 69115 Heidelberg,
Germany. If you have any concerns regarding our products, please
contact ProductSafety@springernature.com

Printed and bound by CPI Group (UK) Ltd, Croydon, CR0 4YY
30/09/2026
02225242-0005